Natural Dam Lake Cuejdel in the Stânişoarei Mountains, Eastern Carpathians

Alin Mihu-Pintilie

Natural Dam Lake Cuejdel in the Stânişoarei Mountains, Eastern Carpathians

A Limnogeographical Study

Springer

Alin Mihu-Pintilie
Interdisciplinary Research
Department—Field Science
"Alexandru Ioan Cuza" University
of Iaşi
Iaşi
Romania

ISBN 978-3-319-77212-7 ISBN 978-3-319-77213-4 (eBook)
https://doi.org/10.1007/978-3-319-77213-4

Library of Congress Control Number: 2018934431

Printed on acid-free paper

This Springer imprint is published by the registered company Springer International Publishing AG part of Springer Nature
The registered company address is: Gewerbestrasse 11, 6330 Cham, Switzerland

Foreword

In Romanian geographic literature, Limnology is a niche science. Limnology studies are artificially segregated by the two fields: Biology or Geography. Both schools avoid an interdisciplinary approach or teamwork. For this reason, the studies that can be highlighted are extremely rare. The same deficiencies are recorded for the study of wetlands, similarly to lakes. The current research direction of the Interdisciplinary Research Department—Field Science, University "Alexandru Ioan Cuza" of Iaşi—is interdisciplinary and transdisciplinary. The study elaborated by the author of this book also follows this direction.

Limnology, as a borderline science, has been avoided for a long time by disciplines designed to deal with understanding interdependent phenomena. Unfortunately, even internationally, the phenomena have not been controlled due to the lack of basic components: the studies are purely biological or purely geographical. Most biological studies take no count of physico-geographical factors. At the same time, geographical studies lack the biological component. We face the inability to fully understand the role of each component and its connections with the whole. In Limnology, there has always been a "missing link."

Specialty schools lead a fierce fight in order to impose their perspective in renowned specialties. How many biologists know what a hydrographic basin is or what its geological constitution comprises? How many specialists take into account the morpho-bathymetric phase as a layering unit for local conditions?

The lithologic substrate, emerged or submerged, represents the development support for communities on various evolution phases. Nevertheless, limnologists talk easily about water chemistry as if it were a gift from God. The fact is omitted that chemical parameters are given by the petrologic composition of hydrographic basin. Through physical and chemical erosion, the elements are detached, transported, and deposited in the lacustrine basin. The lake becomes a depositary for the components of the basin of influence, in this case, the hydrographic basin.

In the Romanian school, as in other European, Asian, or African schools, the lake surface is still confused with that of a wetland itself. Worldwide classifications are often unsatisfactory, and pride cannot be stopped. Fortunately, in science, bribery is not accepted!

The classifications based on wetlands delimitation, according to a series of parameters, have given satisfactory results. From this point of view, the American school defines wetlands based on hydrological, geomorphological, pedological, and biological parameters. The interdependence of the four parameters makes the ambiguity disappear. Nevertheless, the difficulty of delimitation, a long-time research and the high price of analyses interfere.

The link between wetlands and lacustrine basins is extremely strong. Due to this fact, the lake is confused with the typical wetland. Most morpho-hydrographic forms, such as wetlands, are just component parts of water basins or stagnant waters. The most important issue is linked to the interdependence between morpho-bathymetry and limno-ecological parameters existing in an aquatorium.

A mountain unit is highlighted by parameters related to altimetry. Each level of altitude determines a particular type of climate. Climatic parameters, in turn, determine the distribution of plants and animals. A lacustrine basin is the inverted image of a mountain unit. This time, each bathymetric stage corresponds to a number of hydrometric factors. In this case it is about the existence of bathymetric levels. Morpho-bathymetry represents the edifice on which secondary components are installed, exactly like a house waiting for its tenants.

Natural dam lakes from Romania were mainly studied by geographers and rarely by biologists. Most studies are dedicated to the formation process and intend to highlight the evolution in space and time of the water basin. Most of the natural dam lakes from Romania are located in the Eastern Carpathians, where high rainfall and the geological substrate allow landslides. Most landslides were caused by massive deforestation during the late nineteenth century and early twentieth century.

The limno-geographic literature until 2012 demonstrated that the Red Lake is the largest natural dam lake in Romania (12.01 ha). During 2012, topographic measurements were taken for the most important natural dam lakes from Eastern Carpathians and the largest one is actually Cuejdel Lake (13.95 ha). As far as depths are concerned, the same classification is applied: 16.45 m for Cuejdel Lake and 1.05 m for Red Lake. Among the largest natural dam lakes from Romania, the Red Lake is the oldest, formed in 1837. Cuejdel Lake is the youngest, with two formation phases: first, in 1978, when a small lake was formed and second, in 1991, when a massive landslide occurred and the landslide body blocked Cuejdel brook.

As a result of its relative isolation, Cuejdel Lake caught rather late the attention of interdisciplinary research. The lack of modern ways of access leads to its isolation. The entry of Cuejdel Lake into the national and international tourist circuit will be delayed. Unfortunately, most studies focused only on the lacustrine basin and geological substrate and less on the analysis of physical–chemical parameters and water dynamics. Cuejdel Lake is a tourist attraction only for the local population.

This study tries to emphasize the role of morpho-bathymetry in the distribution of sediments and areas of wetlands. At the same time, physical–chemical characteristics and water dynamics are highlighted being directly related to other physico-geographical factors. Thus, the first chapter about natural dam lakes and their position within limnological studies creates an interesting overview about the state of knowledge in the field, both internationally and nationally.

The best represented category on the territory of Romania is one of the natural dam lakes formed due to subsidence and gravitational landslide processes. The author emphasizes that the most favorable conditions for the formation of natural dam lakes are found in the Eastern Carpathians, the Curvature Subcarpathians, and the Moldavian Subcarpathians. These conditions may also occur separately in the Moldavian Plateau. The highest susceptibility is found in the Carpathian flysch area where the formation of lacustrine accumulations behind diluvia is frequent.

The next chapters are focused on methodology and monitoring techniques of Cuejdel Lake. The most modern tools were used in order to take samples, scan the basin, and for integrated analysis. For the first time in Romanian research, the noninvasive technique of bathymetric mapping (GPR) was used. In this regard, as far as the approach and unexpected results are concerned, Chaps. 3, 4, and 5 are distinguished. It is very important that research was conducted based on two aspects: quantity and quality. A meritorious scientific contribution is given by the map of sediment thickness distribution. According to the results, Cuejdel Lake will become clogged in approximately 250 years.

Biotic conditions aimed at describing physico-chemical parameters, water quality index, and trophic level determined based on chemistry and nutrients. In situ measurements were taken seasonally using a Hach Lange multiparameter with real-time display. Laboratory investigations followed standard methods of water analysis. In terms of temperature, Cuejdel Lake behaves like a dimictic aquatorium, with direct thermal stratification in the summer and inverted during winter. The amount of dissolved oxygen decreases inversely proportional to the standard conductivity of water, indicating an euxinic stratification, induced by the lack of vertical currents. The pH value follows the same pattern, with a slightly alkaline environment, specific to mountain waters.

Cuejdel Lake presents a productive environment only in the epilimnion area. The water quality determined based on nutrients, phosphates, and ammonium ions concentration, to which were added to the formula when making calculations the amount of dissolved oxygen, the biochemical demand of oxygen and the pH, indicates a score of 86%. This value is equivalent to first-class quality. Based on the nutrients determined in laboratory samples, Cuejdel Lake belongs to mesotrophic aquatory category. Trophic conditions are similar to those from Lake Sfânta Ana or from Brădişor and Izvorul Muntelui anthropic accumulations.

The trophic level of Cuejdel aquatic limno-system was achieved based on ecological bio-indices determined in the water mass. The following categories were identified: algal communities, vertebrate and invertebrate aquatic fauna, of which we mention insects, zoo-planktonic, and benthic macroinvertebrates communities. Based on ecological bio-indices from the water mass, Cuejdel Lake is defined as an oligotrophic aquatic ecosystem with mesotrophic tendencies during warm season.

At the same time, 11 aquatic and palustrine phytocoenoses were identified which provide favorable habitats for already existing species as well as other new ones. There were reported nine species of birds, one species of amphibians, and one reptile species listed in the Habitats Directive (92/43/CEE) and Birds Directive (79/409/CEE). In this case, having the status of Protected Area (IUCN IV.40 Cuejdel Lake) is not enough for protecting them. Cuejdel Lake area meets all legal conditions to establish a Site of Community Importance (SCI), a Special Area of Conservation (SAC), and/or a Special Protection Area (SPA) according to Natura 2000 methodology.

The entire work is unique as a way of approaching a limnology subject in an interdisciplinary context of European geography. The content of this paper was designed as a novelty in Romanian literature. The author used the latest generation devices, and the interpretation and the results are extremely interesting. Such work brings new data in the analysis of the natural environment and the development of human society in relation to water, the most important resource on Earth.

Iaşi, Romania

Gheorghe Romanescu
"Alexandru Ioan Cuza" University of Iaşi

Acknowledgements

This work was partially supported by the strategic grant POSDRU/159/1.5/S/133391, Project: *Doctoral and Post-doctoral programs of excellence for highly qualified human resources training for research in the field of Life sciences, Environment and Earth Science*, co-financed by the European Social Fund within the Sectorial Operational Program Human Resources Development 2007–2013.

All the equipment used during the data collection process belong to the Geoarchaeology Laboratory within the Faculty of Geography and Geology, University "Alexandru Ioan Cuza" of Iaşi (Manager—Prof. Dr. Gheorghe Romanescu). In this sense, the author thanks Prof. Dr. Victor Spinei, corresponding Member of the Romanian Academy, the director of this research platform, and Conf. Dr. Vasile Cotiugă, the administrative manager of the Interdisciplinary Research Platform ARHEOINVEST.

The author also thanks the scientific coordinator Gheorghe Romanescu, Prof. Dr. at Faculty of Geography and Geology, University "Alexandru Ioan Cuza" of Iaşi, and referents of Ph.D. thesis, who contributed significantly to the consolidation of this scientific endeavor: Senior Researcher Petre Gâştescu at Geography Institute of Romanian Academy and President of Romanian Association of Limnogeography; Prof. Dr. Victor Sorocovschi at Faculty of Geography, University "Babeş-Bolyai" of Cluj-Napoca; Prof. Dr. Florian Stătescu, Dean of the Faculty of Hydrotechnical Engineering, Land Measurement and Environmental Engineering, "Gheorghe Asachi" Technical University of Iaşi.

The commitment of Conf. Dr. Cristian Constantin Stoleriu, during field expeditions and data processing, is highly acknowledged. Also, many thanks to my colleagues from the Faculty of Biology, Geography and Geology and Interdisciplinary Research Department—Field Science, University "Alexandru Ioan Cuza" of Iaşi. Last but not least, I dedicate this work to my family who was worried every time "I disappeared" in the Eastern Carpathians.

Contents

About the Author

Alin Mihu-Pintilie is a Researcher, Dr., at the Interdisciplinary Research Department—Field Science, "Alexandru Ioan Cuza" University of Iaşi (UAIC), Romania. He completed his Ph.D. in Physical Geography, with the Distinction Excellent for his thesis, at the same university in 2015. During his Ph.D. research, he taught university courses in geography, hydrology, and geosciences, an activity that he still undertakes today. Additionally, he has been part of seven national research projects in academic and political institutions in Romania. His main research interests include advanced spatial analysis using GIS software, remote sensing, and geophysics techniques for bathymetric and terrestrial mapping, geoarchaeology, underwater archaeology, and alluvial paleolandscapes. He is the author/co-author of 35 scientific papers and attended 70 scientific presentations and conferences. He is a specialist in continental and oceanic sea diving since 2008 (NAUI Worldwide Patent). In addition, he is a member of national and international scientific societies such as EGEA, ARLG, SOR, SystemIS, GeoDA, and AGR and has also participated in numerous geographical and archaeological expeditions in different parts of Eurasia and the North Atlantic Ocean.

Abbreviations

A	Inferior mesosaprobic area
α–p	Inferior transition meso-polysaprobic area
α_{Global}	Global water flow coefficient
B	Superior mesosaprobic area
β–α	Superior–inferior transition mesosaprobic area
ε_{ave}	Relative permittivity (average)
$\varnothing_{med}$	Grain size (median)
ABA	Siret Water Basin Administration—Romania
AD	Anno Domini
AGR	Romanian Association of Geomorphologists
Aq	Aquitanian (Oligocene–Miocene)
ARHEOINVEST	Interdisciplinary Research in Archaeology—Romania
ARLG	Romanian Association of Limnogeography
ASCE	American Society of Civil Engineers
AT	Anthrosols
BC	Bern Convention, 1979
Bd	Burdigalian (Early Miocene)
BD	Birds Directive
BDO_5	Biochemical Oxygen Demand
C	Critically endangered plants list (sozological category)
C 1…4	Gravity core
Ca^{2+}	Calcium
CDC 401	Standard electrical conductivity of water
Cl^-	Chlorine
CM	Cambisols
CNCSIS	Romanian Council for Research in Higher Education
CS	Cross section
CV	Cumulative lake volume
DEM	Digital elevation model
DF	Dual frequency

DO_2	Dissolved oxygen
E	Endemic plants list (sozological category)
E	Endangered bird species (phenological categories)
E_s	Specific erosion
EGEA	European Geography Association
Eq	Equation
ER	Endemic plant species in Romania
ESRI	Environmental Systems Research Institute
FL	Fluvisols
(g)	Saprobic weight
GeoDA	Geo-speleology Association in Romania
GIS	Geographic information system
GL	Gleysols
GPR	Ground-penetrating radar
GPS	Global positioning system
GSA	Geological Society of America Bulletin
HCO_3^-	Bicarbonate
HD	Habitats Directive
$I_{n\%}$	Average intensity of rain with the probability of n%
ISL	International Symposium on Landslides
IUCN	International Union for Conservation of Nature
$KmnO_4$	Organic matter
LDO 101	Luminescent dissolved oxygen in water
lt-ch	Latorfian–Chattian (Oligocene)
LS	Longitudinal section
LV	Luvisols
NASH	North Atlantic subtropical high
Natura 2000	European Network of Protected Areas
ne-al	Neocomian–Albian (Lower Cretaceous)
Mg^{2+}	Magnesium
mi	Miocene
MN	Minister Order—Romania
MP	Management plan
NH_4^+	Ammonium ions
NO_3^-	Nitrates
NO_2^-	Nitrites
nt	Unthreatening plants (sozological category)
ntR	Unthreatening plant species in Romania
o	Oligosaprobic area
O'_s	Static conductivity
OUG	Government Emergency Ordinance—Romania
p	Polymesosaprobic area
p–α	Inferior polymesosaprobic area
P	Total phosphorus
Pb	Passage bird species (phenological categories)

P 1…7	Water samples
PASP	Specific Alluvial Soil Production
PGA	Peak ground acceleration (seismic activity)
Pg1-lt	Lutetian-Ypresian (Paleocene–Eocene)
PH	Phaeozems
ph	ph of water
Pm	Partially migratory bird species (phenological categories)
PO_4^{3-}	Phosphates
PP-q-E	Rainfall–runoff–erosion equation
pr	Priabonian (Upper Eocene)
Q_{max}	Maximum flow
Q_{avg}	Average flow
Q_{min}	Minimum flow
Q_i	Sub-index calculated for each parameter of water quality
$Q_{n\%}$	Probability of exceeding the maximum discharge (n%)
qh2	Holocene
qp3/3	Upper Pleistocene
qp1/2	Middle Pleistocene
R	Red plants list (sozological category)
RA	Relative lake area
RG	Regosols
ROMSILVA	National Forest Administrations—Romania
ROSCI	Sites of Community Importance Code—Romania
ROSPA	Special Protection Area Code—Romania
RR	Rare plant species in Romania
RRL	Romania Red Lists of bird species
(s)	Saprobic classes
(S)	Saprobic score
S	Sedentary bird species (phenological categories)
Sb	Summer bird species (phenological categories)
SAC	Special Areas of Conservation
SCI	Sites of Community Importance
SGAN	Water Management System of Neamț County
SIL	International Society of Limnology
SystemIS	Association SystemIS Group in Romania
sn-Pg1	Senonian (Upper Cretaceous–Paleocene)
SOR	Romanian Ornithological Society
SPA	Special Protection Area
SOVROLEMN	Romanian-Soviet Society for forest exploitation
STEREO 70	Stereographic Projections for Romania 1970
SO_4^{2-}	Sulfates
T_c	Time of runoff concentration
T°C	Temperature in °Celsius
TDS	Total dissolved salts
TOC	Total organic carbon

USDA	United States Department of Agriculture
USLE	Universal Soil Loss Equation
WQI	Water quality index
UP	Forest Production Units
V	Vulnerable plants list (sozological category)
V	Vulnerable bird species (phenological categories)
vr-tu	Vraconian–Turonian (Cretaceous)
W	Wintering bird species (phenological categories)
W_i	Weight associated for each parameter of water quality
Wr	Wintering rare bird species (phenological categories)
x–α	Xeno-mesosaprobic area

Chapter 1
Introduction

The achievement of a limnological study of a "simple lake" represents a scientific enterprise that is more complex that it appears at first sight. This fact derives from the integrated monitoring concept of an aquatic open system and comes as a reaction to scientific research that has been enforced on modern limnology. Even if the direction of research is geographical, a large part of investigation analysis and methods often concerns an interdisciplinary approach. Taking this into consideration, the results must be treated both in terms of data quantification in a single system and transparently for all other research sub domains that are directly concerned (limnobiology, ecology, hydrology, geomorphology, natural hazards, etc).

Natural dam lakes belong to a more specialized category of lacustrine accumulations, because they present a relatively different character in regards to the methodology of research. These are treated most of the time as a consequence of geomorphologic processes and in some of the cases, their formation and presence represent a risk factor for human society. For this reason, the majority of studies from the international literature belong in the most part to geologists, geomorphologists and geophysicists, and less to limnologists and hydrologists. A very good example from this point of view is represented by the multitude of studies performed within relevant societies, such as: Geological Society of America Bulletin (GSA), International Symposium on Landslides (ISL) or American Society of Civil Engineers (ASCE). Within these specialty reports most of the published information is about natural dam lakes formed over time across the globe.

A remarkable fact is that the natural dam lakes emerge only when a series of direct geomorphologic factors and indirect premises are brought together. The most important of the direct factors are induced by lithologic susceptibility at: slope processes, high lithologic alternation and the layout of permeable rocks package on top of impermeable rocks. The lithologic context corroborates with high declivity and relief fragmentation, intense and long lasting rainfall, valleys with a cross-sectional profile and high slopes, accented lateral erosion induced by rivers, marine or lacustrine abrasion exerted by waves, tides, severe flash floods, floods etc.

The indirect or accidental factors are determined by the seismic activity and volcanism as a trigger factor, lava flows, pyroclastic material accumulation, avalanches and glaciers moving, biogenic activity and the accumulation of vegetation, and so on. In the same category are also included a series of conditions with an indirect char-

A. Mihu-Pintilie, *Natural Dam Lake Cuejdel in the Stânişoarei Mountains, Eastern Carpathians*, https://doi.org/10.1007/978-3-319-77213-4_1

acter, of which probably the most important is induced by massive deforestations which have taken place especially in poor or developing countries.

Costa and Schuster, two American researchers of the US Geological Society, in a paper published in 1987 and revised in 1988, conducted a comprehensive study on typologies and the main characteristics of natural dams that lead to the natural formation of lacustrine deposits. Based on the analysis of over 180 natural dam lakes they identify the fact that: 84% of worldwide natural dams are composed of rocks and materials from landslides and slope processes, 7% have a volcanic origin as a result of the eruptions and lava flows, and 9% present a complex composition and causality. The conclusion of these researchers about the formation phenomenology of natural dams was: "The most frequent dams that block watercourses and determine water accumulation are induced by landslides that mobilize significant amounts of material. Also, they can be the result of the materials transported by rivers during floods, due to gravitational collapse, as a consequence of volcanic activity, eolian processes or can be built by various animals such as beavers" (Costa and Schuster 1988).

Around the world, most of the natural dams formed by landslides occur in mountainous areas, with narrow valleys and steep slopes, where the water accumulation requires a relatively small volume of obstructive materials. Their occurrence is specific to particular areas with an active volcanism or where the ice is the main factor in relief modelling. The temporal dependence of lakes formed in this area is directly proportional to the geomorphologic relationship between natural dams formed and the shape of the valleys. For this reason, within the context of a limnogeographical study where the lake is not separated from its hydrographic basin of origin, the storage of information should be done on a three hydrographic entities level: hydrographic basin, lake hydrographic basin and lake lacustrine basin.

In the mountainous area of Romania, the most favourable conditions for the formation of natural dam lakes are found in the Eastern Carpathians, Curvature and Moldavian Subcarpathians, but also in some isolated areas of the Moldavian Plateau. The highest susceptibility occurs in the Carpathian flysch area, where the layers of permeable and impermeable sequences, the lithologic alternation with different hardness deposits and the accented declivity of relief trigger slope processes manifestation and in some of the cases the formation of lacustrine accumulation.

Natural dam lake Cuejdel represents an almost unique hydro-geomorphological entity in the landscape of the Eastern Carpathians, because the reduced distribution of accumulations formed by diluvial dams depends on their ephemeral nature. There are many examples in the specialty literature that confirm lakes with a short existence in the landscape and negative hydrologic events caused by their destruction (floods and flash floods). The permanence of this lake within the Stânișoarei Mountains landscape for a longer period of time (25 years), created the premise for a limnogeographical study to confirm or disprove the aquatorium stability against environmental factors. In Romania, a similar study was conducted only on Red Lake, formed in 1837, on Bicaz Valley—Hășmașul Mare Massif (Romanescu et al. 2013).

Before the landscape changes that occur on Cuejdel Valley, the inhabitants stipulated to some extent the transformations that would take place in the superior basin, upstream Gârcina (Neamț County). Nevertheless, even though a series of landslides have occurred in the area with a tendency to obstruct the valley, the secondary stage from 1991, during which the present Cuejdel Lake would have formed, was not described. Although a rarity in 1992, the Romanian geographer Ichim and his co-workers, achieved the first bathymetric map, attesting the formation of the largest natural dam lake in Eastern Carpathians (Ichim et al. 1996).

According to the monographic attestation of lacustrine basin genesis, described by a local and a Geography teacher at the same time, this was due to "Particularly abundant rainfall by the end of July—334.8 mm, of which 130 mm only in one day, added to over 100 mm recorded at the beginning of August, have saturated the crumbly rock from the western and southern versant of Muncel Mountain, which suffered a massive landslide, stopping in the valley of Cuejdel brook" (Baroi 2012). Subsequently, the name of Cuejdel Lake was given after the hydronym of the brook on which it formed, this coming from Slavonic (Chivejdi), which means "dry valley" or "river stone". It is also known as the Crucii Lake (Cross Lake) after the toponym Piciorul Crucii (Cross "Foot"), which is situated on the right versant of the lake.

In this context, the general objective of the text is to provide an interdisciplinary study of the Cuejdel lacustrine basin, by triple integration of the investigation means specific to limnogeographical research: the analysis of the physical–geographical background (Cuejdiu hydrographic basin, Cuejdel lake hydrographic basin), lacustrine biotope (lake basin, sediments and water) and aquatic and palustrine biocoenoses (vegetation and fauna). The specific objectives of the study are:

- To identify the physical-geographical factors that caused the slope processes and diluvial obstruction of the Cuejdel brook;
- The analysis of hydro-geomorphological balance that characterizes the present natural dam and accumulated water volume (the evaluation of hydrologic risk);
- To describe the variation of main morphometric and morpho-bathymetric elements during the 25 years since the lake formation (water surface, maximum depth, levels, water volume, lacustrine basin profile, etc.);
- To estimate the clogging rate and indentifying the sediment budget through direct and indirect measurements according to the equations of erosion determination;
- The analysis of biotic conditions within the aquatic ecosystem, regarding the description of physical-chemical parameters, water quality index and trophicity condition determined based on the chemistry of the water and ecological bio-indices;
- The description of aquatic and palustrine phytocoenoses, vertebrate and invertebrate fauna, as well as the potential of natural habitats in sustaining an avifaunistic and aquatic biodiversity.

This study is considered to form a useful database for the environmental authorities licensed to reach important verdicts regarding the preservation of natural dam lake Cuejdel. Also, the spatial database can be used in hydrotechnical–development projects intended to minimize the negative fluctuations of water levels, drainage channel erosion, shore and natural dam strengthening.

References

Baroi C (2012) Gârcina, pagini de monografie (Gârcina, monographic pages). Cetatea Doamnei Publishing House, Piatra Neamţ

Costa JE, Schuster RL (1987) The formation and failure of natural dams (No. 87-392). Geol Soc Am Bull. US Geological Survey Report, Vancouver, Washington

Costa JE, Schuster RL (1988) The formation and failure of natural dams, 2nd edn. Geol Soc Am Bull 100(7):1054–1068

Ichim I, Rădoane N, Rădoane M (1996) Procese geomorfologice cu intervale de recurenţă mare în arealul munţilor flişului. Exemplificări din Judeţul Neamţ (Geomorphological process with high frequences in flysch mountains area. Examples from Neamț County). In: Studii şi cercetări—extras 8, Muzeul de Ştiinţe Naturale Publishing, Piatra-Neamţ, p 15

Romanescu G, Stoleriu C, Enea A (2013) Limnology of the Red lake, Romania. An interdisciplinary study. Springer, Dordrecht, Heidelberg, New York, London

Chapter 2
Natural Dam Lakes and Their Status Within Limnological and Geographical Studies

Abstract According to the geographical terminology, Limnology is a department of science that studies natural and artificial lakes taking into account their formation, hydrological, thermal and chemical conditions, the developmental capacity of flora and fauna associations, the connections to the environment and the potential for exploitation by man. Natural dam lakes are included in a rather special category of lacustrine basins, since they display a different set of characteristics as far as the research methodology is concerned. On most occasions they are regarded as a consequence of geomorphological processes and even as a hazard factor for human society. For this reason, the majority of the studies in the international literature belong to geologists, geomorphologists and geophysicists, with a significantly low percentage of them belonging to researchers in other fields. In accord with the viewpoint underlying the present study, we consider that the Cuejdel lacustrine basin, at present, following all the premises accepted by both sides (geographers and biologists) fits into the definition of a natural lake. The Red Lake is perhaps the best known natural dam lake in the Eastern Carpathians. Until recently it had been considered the largest natural dam lake in Romania, surface-wise as well as volumetrically, but after 1991, when the natural dam lake Cuejdel was formed (Stânişoarei Mountains), it moved to the second position in this classification. The work "Limnology of the Red Lake, Romania. An interdisciplinary study", authored by Romanescu et al. (2013) and published by Springer, basically acted as a model for the present study, which is meant as a continuation of the research on the natural dam lakes formed in the Romanian Carpathians.

2.1 The Interdisciplinary Character of Limnological Studies

Together with the assertion of limnology as a subfield of research amidst hydrosciences, a scientific debate emerged on its affiliation, as a secondary branch, to either Biology or Geography. For this reason, in the speciality literature, one can find a series of definitions describing the field of Limnology in different terms, based on the scientific background of the researchers emitting them. We believe that the

A. Mihu-Pintilie, *Natural Dam Lake Cuejdel in the Stânişoarei Mountains, Eastern Carpathians*, https://doi.org/10.1007/978-3-319-77213-4_2

review of these different perspectives is useful in underlining the interdisciplinary character of the research of lacustrine formations and implicitly of the present study.

From an etymological point of view, the word "limnology" has its roots in the Greek limné—(lake), a term defining the catchment or accumulation of water in formations like lakes or ponds, as well as in minor or micro-forms of basins, known as limnology-related dialectal terms: mountain lake, tarn, pond, pool, dug-out etc., to which the termination "logos" (meaning "study") is added. From the very first treatises of general physical geography, as well as the papers regarding lakes, there is a specification about limnology studying mostly lakes or lacustrine accumulations (Halbfass 1923; Martonne 1925; Hutchinson 1957; Kalesnik 1956).

According to the Romanian geographical terminology, "Limnology is a department of science that studies natural and artificial lakes taking into account their formation, hydrological, thermal and chemical conditions, the developmental capacity of flora and fauna associations, the connections to the environment and the potential of exploitation by man" (Gâştescu 1998). In a revised version, "Limnology is the science that studies lakes and all the other forms of stagnant water (natural or artificial), including hydrological, physical and chemical phenomena, in relation to the natural environment (from a chemical, thermic and dynamic point of view, as well as the capacity of development of the flora and fauna, etc.), as well as the potential of exploitation by man" (Romanescu 2008).

From the perspective of biologists and particularly hydrobiologists, "Limnology is the science that studies continental waters from a biological standpoint, comprising running and stagnant waters, springs, irrespective of their nature (sweet, salty or briny), consequently including the study of the hydrosphere in Oceanography and Limnology" (Goldman and Horne 1983; Gâştescu 1998).

Due to its focal interest, limnology is a borderline field of study, situated at the interface between Geography and Biology, closely connected to other sciences from which it borrows certain research methods. This interdisciplinary feature of its investigation and analysis methodology of lacustrine exploration has sparked, throughout its existence, a series of debates regarding its affiliation, either as a sub-branch of Hydrology or Biology, or as a component of Geography.

The study of lacustrine formations belongs to Geography as it explores their depression and manner of formation, the accumulated water volume and their physical, chemical and hydro-dynamic characteristics, the degree of vulnerability, the risk potential and so on. Likewise, limnology belongs to Biology, since its inquiries also focus on the flora and fauna in standing waters, as well as in the areas under the immediate influence of lacustrine formations (benthal area, damp areas, wetlands and others). In other words, the lacustrine biotope (the lacustrine depression and the water) represents a sub-branch of limnology, called physical limnology or the geography of lakes, while bio-hydrocoenosis (lacustrine biocoenosis) represents a different one, called biological limnology, or the biology of lakes (Gâştescu 1979; Romanescu 2008).

As the biotope and bio-hydrocoenosis of a lake function as a whole, also called a limnosystem or a lacustrine ecosystem, the fields of expertise for geographers and biologists get interconnected in the realm of limnology, completing each other.

Grounding his assumptions on the systemic theory, reaffirming the interdisciplinary character of limnology, Hutchinson (1957) stated the following in his limnology treatise: "Lakes, for that matter, form systems which are more or less closed, so that they exhibit an entire range of ecological environments which allow for an approach to nature's mechanism based on a real comparison".

In a more comprising vision, in order to avoid the conceptual gap between the scientific approaches belonging to fields of study close to Limnology, the latter can also be defined as an "interdisciplinary science focusing on the water basins—natural or artificial—with slow exchange, from a historical standpoint, meaning dynamically in time and regionally as space, the current physical and biotic processes, against the background of transforming energy and substance within the water mass, the laws governing evolution and geographic distribution, natural resources and the possibilities of exploitation for the benefit of the human society" (Gâştescu 1971).

If from a methodological standpoint one can eventually notice common conceptual ground between geographers and biologists, opinions do begin to differ when it comes to defining specifically limnological terms. There are several points of view addressing the fundamental notions, among which we mention "lake", "reservoir" or "pond", definitions issued by different limnology schools, where they were shaped in order to serve their own research purposes (biology, geography, chemistry and so on). For the present study, eliminating uncertainty at least from a geographical standpoint is useful for an adequate approach to modern limnology.

The first references define a lake in a rather sketchy manner as being: "… a body of still water, in a basin, without a direct connection to the sea" (Forél 1892); "We shall adhere to the term of lake for all natural or artificial basins containing water and their own small-scale ecosystems, with a certain degree of autonomy…" (Dussart 1966); "A lake is the body of water which is relatively stagnant, kept in a basin on the Earth's surface, with no direct connection to the Planetary Ocean" (Gâştescu 1971).

At that time, resorting to these definitions alone, one could not draw a clear distinction between lake, reservoir or pond. The transition of the categories of lakes remained ambiguous until it was decided, in a generalised manner, that distinguishing them according to bathymetric criteria as Imboden (1976) did—supported by the majority of the Romanian limnologists—is satisfactory, to a certain extent, for the need of adequate delineations. Imboden stated that "… a lake displays, besides a coastal area, a deep area devoid of light, where the bottom of the lake is denuded of vegetation; its water is often stagnant as currents are active especially when the lakes has major affluents or outlets" (Imboden 1976).

The Romanian limnobiologist Antonescu (1967), inferring this manner of characterizing lakes, also states that: "We call a lake a stagnant water which is wide and deep enough as to allow us to distinguish two areas within it: (1) A strand area (bank) covered with emersed and submersed aquatic vegetation, so with a productive bottom and (2) A deep (profound) zone, obscure, deprived of plants and with the unproductive bottom covered in finely-textured, uniformly structured sludge" (Antonescu 1967).

From a geographical viewpoint, its relatively stagnant character is accepted in the definition of the lake because it sets it apart from the river. As far as the delineation

Fig. 2.1 The natural dam lake Cuejdel–Stânişoarei Mts. (July 2012)

of lakes based on surface and depth indicators are concerned, one considers that the previous definitions possess an acutely biological character and does not contribute to the clarification of the concept of "lake". As an argument, one usually resorts to the example of the Azov Sea which, with a depth of 9 m and lacking deep water per se, obscure and unproductive, is to be included in the category of puddles according to the classification issued by biologists. Geographers consider that one of the most important criteria of differentiation between lakes and other lacustrine formations is the relation and direct position to the nearest marine basin. Concerning this, "…it is to be underlined that the lake has no direct connection to the sea and we must comprehend that the lake and the sea are not situated at the same level, so that they cannot influence each other, as in the case of marine bays. Lagoons, found near the sea shore, are still included in the category of lakes" (Gâştescu 1998).

In accord with the viewpoint underlying the present study, we consider that the Cuejdel lacustrine basin, at present, all the premises, accepted by both sides (geographers and biologists) to fit into the definition of a lake. This statement can be sustained by the fact that the water basin is characterized by a slow exchange with the surrounding aquatic environments, has a differentiated ecosystem, with its own corresponding fauna, it is defined by a shoreline covered in vegetation, with a productive lacustrine bottom, and a deep water area which is obscure and denuded of vegetation, while the water mass is stagnant, registering significant level oscillations only when its main emissaries encounter massive increases in flow (Fig. 2.1).

Besides the aspects mentioned above, as the present analysis resonates conceptually with the guidelines of limnogeographical research, we consider that the natural dam lake Cuejdel is a constituent of the hydrographic basin of the river Cuejdiu, with direct implications at the level of drainage. This approach fences off the representation of the lacustrine unit as a closed system, unitary from the viewpoint of the geographical landscape in which it is included, an aspect which draws us away, to a certain extent, from the biologists' analytical approach. However, we shall attempt to integrate the investigation and monitoring techniques linked to the lacustrine environment in an interdisciplinary perspective, as a complex limnological study is to be based on such a methodological model.

2.2 A Short Description of the Evolution of Limnology Within the Field of Hydro-sciences

Beginning with ancient times, the human communities have used the water stored in the natural lakes for the supply of drinking water and irrigations. Simultaneous with technological development and consequently of water consumption, humanity began to set up lakes for much more diverse purposes. The oldest mention of creating a basin through building a dam on a river is of the Kosheish dam (about 2900 BC), on the Nile, downstream from Memphis, under the reign of Pharaoh Menes, the founder of the first Egyptian dynasty (Gâştescu 1979; Plăvan 2010).

From Aristotle's first works and continuing with Ptolemaeus's writings we have known that the water accumulated in the lacustrine formations has played a major part in the development of the humankind. At the beginning of the first millennium AD, in his work "The Natural History", Pliny the Elder makes a synthesis of all knowledge regarding life forms in waters. Even if the times of the Middle Ages are characterized by limited input in the field of nature studies, during the Rennaissance one can notice a renewed interest in observing the natural environment, based on the knowledge inherited from antiquity, an interest materialized through desultory studies with references to the species of fish in lakes.

Even if human society has always manifested a desire for more in-depth knowledge of lacustrine accumulations, albeit natural or artificial, it would not be until the end of the XIXth century that the study of such formations would benefit from a truly scientific component. For this reason, the founder of limnology as a branch of science and not as a way of lakes utilization is François-Alphonse Forel (1841–1912), medical doctor and natural historian, Professor at the University of Lausanne (Switzerland). His first work, "Matériaux pour servir à l'étude de la faune profonde du lac Léman" (Forél 1869), was to open the way for ample monographies of this lake, published in three volumes "Le Lémane-monographie limnologique" (Forél 1892, 1895, 1904). Together with another limnology workbook, titled "Handbuch der Seenkunde: allgemeine Limnologie" (Forél 1901), published in Stuttgart, and 100 other papers, the whole collection is considered to be the scientific basis for the rather delayed debut of limnology as a distinct science within hydro-sciences (Romanescu et al. 2013).

Subsequently, numerous researchers from other fields (physicists, chemists, geographers, geologists, biologists, etc.) contributed to the development of limnology, through their studies on the lacustrine environment. One of them is James Chumley, who, in 1910, gathered an appendix of more than 2900 titles and notes for the book written by Murray and Pullar (1910) about Scottish sweet water lakes, published in Edinburgh. Thus, due to his exhaustive studies on Scotland's lakes, Sir John Murray is regarded as the first limnologist who introduced the methodology of oceanographers and hydrologists into research on lakes (Gâştescu 1998).

Another publication which put its mark on the research of lacustrine formations at a global level was "Grundzüge einer vergleichenden Seenkunde", issued in Berlin and edited by Halbfass (1923). The work functioned as a basis for classic limnology until the issue of the first volume of the series "A Treatise on Limnology-Geography, Physics and Chemistry"—Hutchinson (1957). The editions to follow further fathomed the research methodology, establishing the status of limnology within biosciences: "Introduction to Lake Biology and the Limnoplankton"; "Limnological Botany"; "The Zoobenthos" (Hutchinson 1967, 1975, 1993).

Besides redesigning conceptually the systematic methodology of analyzing the biotic environment of lakes, the importance of the research carried out by both Halbfass and Hutchinson (particularly the latter) also lies in accomplishing a detailed classification of lacustrine formations based on the genesis of the basin/depression as a form of relief, as well as from the viewpoint of the geomorphological process which added them to the landscape. The completeness of this classification resides in the discovery of 76 types of formative processes, included in 11 classes of natural processes leading to the appearance of lacustrine formations. At that date, the new taxonomy determined a unanimous acceptance from the part of the international scientific community, as well as Romanian limnologists.

Another important event took place on the 1st of January 1922, in Kiel–Germany, with the formation of the International Society of Limnology (SIL), on the initiative expressed by the German zoologist Thienemann and the Swedish botanist Naumann, supported by 103 researchers from other countries. This organization was formed as a response to the increasing number of researchers in the field of limnology on all continents, and remains active up to the present day.

Consequent to the founding of SIL, most of the publications prominent in the field of limnology, at a global level, were the result of the work done by researchers from SIL. Among them, the majority had basic training as biologists, but there existed a number of limnologists whose studies focused especially on lacustrine geomorphology (Ruttner 1963; Cole 1968, 1994; Smith 1975). From the category of geomorphologists, we can mention Pouriot and Meybeck, who in 1995 established a new direction of research in modern limnology, according to which the lake establishes functional relationships with the entire drainage basin and, accordingly, the lacustrine analysis has to be extended beyond the lake itself, as in the case of the wetlands adjacent to aquatic formations (Pouriot and Meybeck 1995; Tiner 1999).

In 2010, Menshutkin and Levchenko, two Russian statisticians, elaborated an analysis model for developing limnology as a science. Based on the number of limnologists, most of them active members of SIL, as well as on the algorithm describing the accumulation of scientific information in the field of limnology over a period of 100 years (1900–2000), they identified the main stages of massive information build-

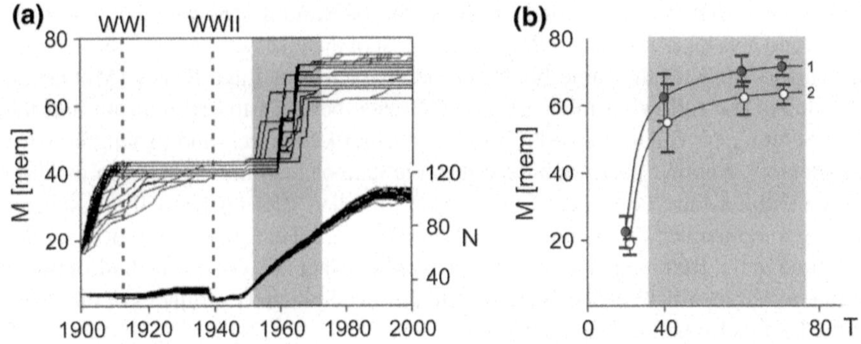

Fig. 2.2 **a** An increase of the information filling of limnology M (mem) and a change of the number of limnologists N in model experiments for the time interval from 1900 to 2000; **b** Dependence of the information filling of limnology M (mem) on the mean lifespan duration of scientists T (age) for the time interval from 1900 to 2000: (1) Only the accumulative mechanism of the appearance of scientific discoveries; (2) Only the intuitive preseeing mechanism of the appearance of discoveries (Menshutkin and Levchenko 2010)

up. Similarly, they identified the linear periods, when no scientifically notable events were registered in the field (Menshutkin and Levchenko 2010).

The importance of this study is augmented by its chronological description of the main research directions, induced by the appearance of benchmark limnological studies. Likewise, the interconnectedness between the build-up of scientific information and the life expectancy of limnologists is well documented, as in the case of the intuitive mechanism of the pressure imposed by scientific discoveries or the improvement of technology. At present, at a global level, one can notice an increase in the number of studies dedicated to lakes, due to the multitude of sub-branches of limnology, which implies the active involvement of researchers from various fields. This fact is particularly noted in the international specialist literature, where case-studies and applications related to lacustrine formations are issued each year by biologists, chemists, geographers, geologists, physicists, hydro-technicians and so on (Fig. 2.2).

In Romania, the first information on lakes is to be found in various historical documents, being connected to the concessions of domains comprising, among others, lakes, reservoirs and ponds. In the historical documents of Wallachia up to the year 1600, numerous lakes are mentioned, most of them in the everglades of the Danube and in Oltenia (in the South of present-day Romania). Some lakes in the Transylvanian Depression are known from the Roman period. The ones in the Transylvanian Plain are mentioned in various documents in the XVth century; thus, we find out that 13 lakes arranged in a cascade were to be found in the Valley of Luduş (Gâştescu 1963).

Information comprising basic scientific explanations is included in "Descriptio Moldavie", authored by the Romanian scholar Dimitrie Cantemir in the first half of the XVIIIth century. Here, he mentions that within the bounds of the medieval

Moldavia (Eastern Romania and the Republic of Moldova, nowadays) "there are rivers, and also lakes (ponds), in Moldavia. Most of the lakes are partly natural, partly man-made". It is in the same book that he characterizes Lake Brateş (Moldavian Plateau) in the following manner: "Lake Brateş, between the river Prutu and the river Seretu, near Galaţi; its width is one and a half kilometres, and its length is two kilometres". Another paragraph describes the lake of Dorohoi, situated in the Jijia Valley (Moldavian Plain), next to the spring of Jijia: "Not far from Jijia's source, worthy to remember for its multitude of fish species". Thus, one can notice, for the first time in the Romanian natural history, a more exact analysis of the hydrological regimen characterizing the lakes situated in the meadows and everglades of the rivers in the East of Romania (Gâştescu 1963).

The first scientific studies of lakes in Romania were issued in the second half of the XIXth century and they focused primarily on the salt-water lakes at the seaside (the shore of the Black Sea) and in the Romanian Plain. Among the authors we shall mention here Antipa (1910), Murgoci (1907), Brătescu (1915), Mihăilescu (1924) and Morariu (1946). The first studies of the chemistry of lakes and the therapeutic properties of salty lakes are also worth mentioning—they were authored by Romanian doctors and geologists such as Ştefănescu (1888), Mrazec (1901), Maxim (1929) and Petrescu (1932). We can also mention studies of the economic importance of lakes, done by Bujor (1902), Antipa (1916), Borcea (1926), Antonescu (1932) and Băcescu (1942). With regard to the morphogenetic analysis of lacustrine basins and subsequently the classification of lakes on the Romanian territory, we shall mention Brătescu (1915), Mihăilescu (1940), Orghidan (1932), Năstase (1949) and Şenchea (1948).

A benchmark study for limnology is "Die Seen Rumäniens und ihre Bewirtschaftung" by Lepşi (1935), published in Chişinău, in which the author does the first classification of the lakes in Romania based on their formation. After the publication of the book, the classification was to be considered incomplete, as it had failed to include references to lakes formed by compaction, carst and wind action, natural dam lakes or lakes formed by landslides. The completion of the taxonomy associated with the formation of lakes was to be done by Mihăilescu (1932), comprising the entire array of lake types, grouping them according to the Romanian forms of relief; his work was followed by Şenchea's (1948), who classified lakes based on their chemical composition.

After 1950, the limnological studies in our country intensified, with authors reissuing the matter of the river and sea firths, the first being Gâştescu (1959), followed by Rădulescu and Năstase (1960). It was in the same period that the lakes in the mountaneous parts of Romania (Făgăraş, Meridional Carpathians) became the focus of intensive research, the most important information being synthesized in the studies by Pişotă (1956, 1957, 1958, 1971), Năstase (1960) and Mălăcea (1956).

Based on the accumulation of hydrometric information from numerous papers published in the field of limnology, particularly on water evaporation at the surface of lakes, the first works of a more holistic type appeared, written by Ujvari (1959) and by Ujvari and Gâştescu (1958). In Romanian scientific literature, among the most significant works on regional limnology with a notable geographic character include the following monographs: "Câteva aspecte privind bilanţul hidrologic al lacurilor din Câmpia Moldovei" (Schram 1970), "Lacurile dulci din Câmpia Transilvaniei" (Săndulache 1970), "Lacurile din România" (Gâştescu 1971), "Lacurile glaciare din Carpaţii Meridionali" (Pişotă 1971), "Lacurile de pe litoralul românesc al Mării Negre" (Breier1978).

After 1990, the ample limnological works suffered a reduction in numbers, the speciality literature being enriched with revised editions of older treatises of limnology (Gâştescu 1998; Romanescu 2008; Romanescu et al. 2013). However, an important breakthrough for limnological research was noted in 2008, when the Romanian Association of Limnogeography (ARLG) was founded, having as founding members the most renowned limnogeographers in the major academic centres of the country. The association's publication, titled "Lakes, reservoirs and ponds—Romanian Journal of Limnology" represents, at present, the most valuable source of information regarding the research of lakes in Romania.

2.3 Natural Dam Lakes in International Literature

Natural dam lakes are included in a rather special category of lacustrine basins, since they display a different set of characteristics as far as the research methodology is concerned. On most occasions they are regarded as a consequence of geomorphological processes and even as a hazard factor for the human society. For this reason, the majority of the studies in the international literature belong to geologists, geomorphologists and geophysicists, with a significantly low percentage of them belonging to researchers in other fields.

A really good example for this is given by the multitude of the studies and analyses carried out within specialised societies, such as: Geological Society of America Bulletin (GSA), International Symposium on Landslides (SIL), or American Society of Civil Engineers (ASCE). Their conferences and other events provided an adequate stage for most of the dissemination of information regarding natural dam lakes formed throughout time all across the globe (Davis 1882; Alden 1928; Adams 1981; Code and Sirhindi 1986; Li et al. 1986; Costa and Schuster 1987; Meyer and Leidecker 1999).

Observing the above-mentioned studies, one can notice the fact that natural dam lakes emerge in the landscape only when certain direct morphological factors and a series of indirect premises exist. The most important direct factors are induced by the lithological susceptibility to various geomorphological processes on the slopes, the high degree of lithological alternation and the layout of the layers of permeable and impermeable rocks. The lithological context is colligated to high declivity and the fragmentation of relief, long-term heavy rainfall, valleys with steep slopes, pronounced lateral erosion caused by rivers at the foothills, lacustrine or marine abrasion triggered by waves or tide, significant high waters and floods etc. (Costa and Schuster 1987, 1988; Meyer and Leidecker 1999; Noord and Phuntsho 2005; Schneider et al. 2011; Romanescu et al. 2013) (Fig. 2.3).

Fig. 2.3 Photos-document evidence of some landslide dams from around the world: **a** A landslide caused by an earthquake in 1911, in Central Asia (Tadjikistan), on Murghab River (Pamir Plateau), forming Sarez Lake; **b** upstream view along the Min River, Sichuan Province (China), at the remains of the 1933 earthquake-induced Yinping landslide dam, which impounded Da Lake (middle background), Xiao Lake (foreground) was impounded by another landslide dam caused by same earthquake; **c** the landslide dam formed by the 1959 Madison Canyon landslide, Montana (USA), and of Earthquake Lake, wich it impounds. Photograph by JR Stacy—US Geological Survey) (Costa and Schuster 1988)

Indirect or accidental factors are determined by the seismic activity and volcanic activity as trigger-factors, lava discharches, the accumulation of pyroclastics materials, avalanches and the movements of glaciers, biogenous activity, the growth of vegetation etc. Also included here is a series of indirect conditions, probably the most important one is caused by the significant degree of deforestation observed particularly in poor or developing countries (Chang 1984; Wu 1986; Costa and Schuster 1987, 1988; Schneider et al. 2011).

Two American researchers from the US Geological Society, Costa and Schuster (1987, 1988), completed an ample work on the typologies and main characteristics of natural dams which lead to the emergence of lacustrine formations naturally. Following the analysis of more than 180 natural dam lakes, they showed that 84% of all natural dam lakes in the world are formed by rocks and materials produced by landslides and slope processes, 7% have a volcanic origin, being formed through seismic eruptions and lava discharge, and 9% display complex composition and causality. The same researchers state that "The most commonly encountered dams obstructing watercourses and determining water accumulation are the effect of landslides, which transport massive quantities of sediments. Also, they can be formed by the sediments transported by rivers during floods, due to gravitational collapses and caving in, as a result of volcanic activity, wind processes, or can be constructed by various animals, like beavers" (Costa and Schuster 1987).

2.3.1 *Classification and Typology of Natural Dam Lakes*

The classification of natural dam lakes has been done by several researchers throughout time. The American geologist Davis (1882) was among the first to notice a certain degree of differentiation at the level of genesis, typology, formation process and a type of accumulation of water by means of natural dams. From this perspective, another important researcher is the British limnogeographer Hutchinson (1957), who introduced in the general classification of lakes a category named "Natural dam lakes", identifying three other sub-types: (1) Lakes formed by natural dams after landslides, mud flows and debris flows; (2) Lakes formed in-between the alluvial mass and a flank of a versant; (3) Lakes formed after valleys are dammed by trails of debris.

The most complete classification of natural dam lakes is provided by Costa and Schuster (1987), who succeed in merging, in an original manner, all the classifications done by that time by other researchers. Next to the well-established categories, they manage to add a series of novel sub-types, like eolian dam lakes, lakes formed in coastal areas, lakes formed by the activity of animals etc. (Table 2.1).

Most of the natural dam lakes formed as a result of landsliding are seen in mountainous areas, with narrow valleys and steep slopes, where the accumulation of water needs a relatively low volume of obstructive materials. They are particularly frequent in areas with active volcanoes or in areas where ice dominates as far as modelling the relief is concerned. The temporal interdependence of the lakes formed in this area is in direct ratio to the geomorphological relationship between the natural dams formed there and the shape of the valleys. Thus, several types of natural dams have been identified, corresponding to a certain degree of populating the valleys, classified by Swanson et al. (1985) in 6 distinct categories (Evans 1984; Costa and Schuster 1987) (Table 2.2).

The lakes in the first category are neither deep nor wide, and most of the times they cause damage only if the dams is deteriorated, and the accumulated water floods the downstream area. The Type II lakes are larger and more dangerous than the ones in the first category, because the mass of accumulated water is increased (Swanson et al. 1985). As far as Type III is concerned, the lakes become even more dangerous,

Table 2.1 Types of obstructive dams—most dangerous dams are underlined (Romanescu et al. 2013; Costa and Schuster 1991; modified from Hutchinson 1957; Schuster and Costa 1986)

Type and subtype of dam	Example of natural dam lakes with references
1. Volcanic dams	
1.1 Volcanic peaks	• Nicaragua Lake, Nicaragua (Hutchinson 1957)
1.2 Lava flows	• Sang Lake, California (Finch 1937)
<u>1.3 Pyroclastic flow</u>	• Rio Magdalena Lake, Mexico (Silva et al. 1982)
2. Landslide dams	
<u>2.1 Slides, slumps</u>	• Earthquake Lake, Montana (Hadley 1964)
<u>2.2 Mud, debris, earth flows</u>	• San Cristobal Lake, Colorado (Crandell and Varnes 1961)
<u>2.3 Rock, debris avalanches</u>	• Spirit Lake, Washington (Meyer et al. 1986)
2.4 Liquefaction of sensitive clays	• Yamaska River Lake, Quebec, Canada (Clark 1947; Costa and Schuster 1991)
2.5 Peat slides	• Addergoole Bog Lake, Ireland (Costa and Schuster 1991)
2.6 Scree	• Goatswater Lake, United Kingdom (Marr 1916)
3. Glacial dams	
3.1 Ice	• Gapshan Lake, Pakistan (Mason 1929)
<u>3.2 Moraine</u>	• Nostetuko Lake, British Columbia, Canada (Blown and Church 1985)
3.3 Ice and snow avalanche	• Rio Plomo Lake, Argentina (King 1934)
4. Fluviatile dams	
4.1 Tributary sediments	• Pepin Lake, Minnesota-Wisconsin (Davis 1882)
4.2 Main-channel sediments, (lateral lakes)	• Tung-ting Lake, China (Hutchinson 1957)
4.3 Alluvial fans	• Tulare Lake, California (Hutchinson 1957)
4.4 Deltas	• Blue Lakes, California (Davis 1933)
4.5 Levee deposits (oxbow lakes)	• Old River Lake, Louisiana (Costa and Schuster 1991)
5. Eolian dams	
5.1 Dunes	• Moses Lake, Washington (Costa and Schuster 1991)
6. Coastal dams	
6.1 Bay-bars	• Freshwater Lagoon, Eureka, California (Cotton 1941; Costa and Schuster 1991)
7. Organic dams	
7.1 Logs and other vegetation	• Okeechobee Lake, Florida (Hutchinson 1957)
7.2 Beaver dams	• Beaver Lake, Montana (Costa and Schuster 1991)

Table 2.2 Classification of landslide dams

Type	Diagrams of mass-movement processes most likely to form particular landslide dams	Landslide dams geomorphically classified with respect to their relations with the valley floor
I. Falls slumps	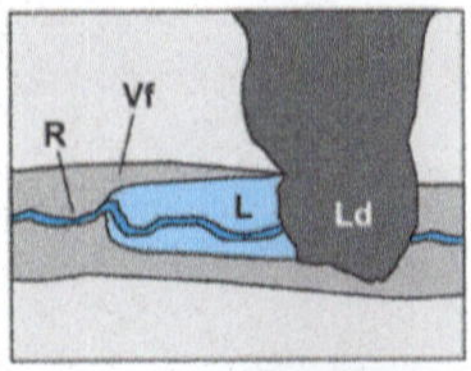	Dams are small with respect to the width of the valley floor and do not reach from one valley side to the other
II. Avalanches slumps/slides	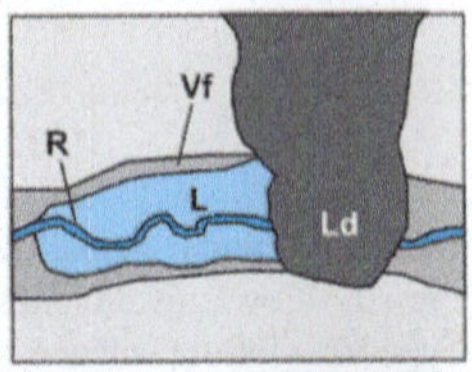	Dams are larger and span the entire valley floor, occasionally depositing material high up on opposite valley sides
III. Flows avalanches	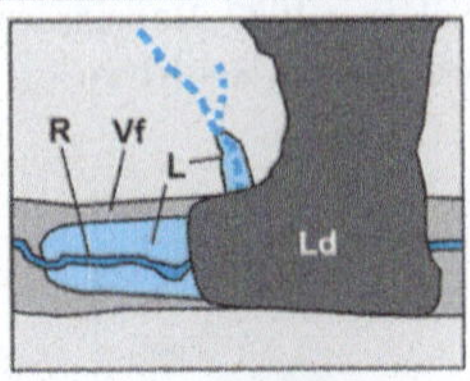	Dams fill the valley from side to side and move considerable distances upvalley and downvalley from the failure. These dams typically involve the largest volume of landslide material
IV. Falls slumps/slides avalanches	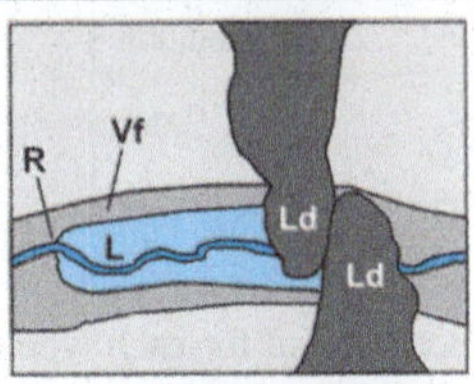	Landslide dams form by the contemporaneous failure of material from both sides of a valley. The landslides can adjoin head-to-head in the middle of the valley, or they can juxtapose one another
V. Falls avalanches	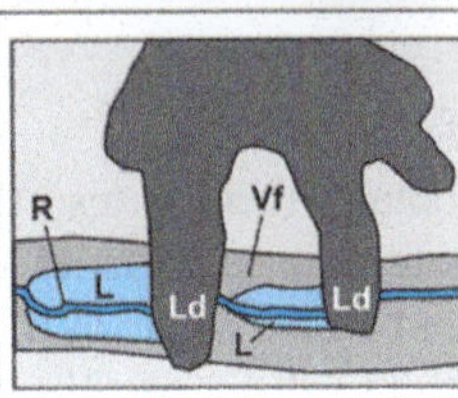	Landslide dams form when the same landslide has multiple lobes of debris that extend across a valley floor and form two or more landslide dams in the same reach of river

(continued)

Table 2.2 (continued)

Type	Diagrams of mass-movement processes most likely to form particular landslide dams	Landslide dams geomorphically classified with respect to their relations with the valley floor
VI. Slumps/slides	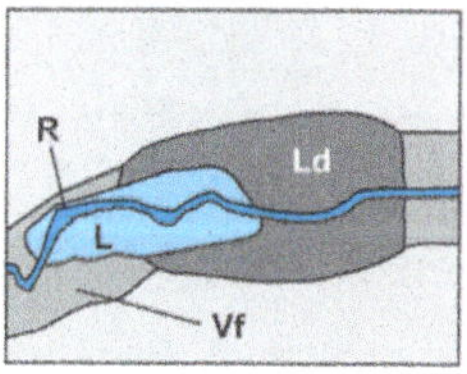	Landslide dams involve one or more failure surfaces that extend under the stream or river valley and emerge on the opposite valley side from the landslide. These dams typically involve slow basal sliding and slumping and form lakes by raising the elevation of the streambed, changing the local gradient of the stream

Mass-movement processes most likely to form particular landslide dams are: Type I—falls, slumps; Type II—avalanches, slumps/slides; Type III—flows, avalanches; Type IV—falls, slumps/slides, avalanches; Type V—falls, avalanches; Type VI—failure surface, slumps/slides (Clark and Waldron 1984; Swanson et al. 1986; Costa and Schuster 1987; Romanescu et al. 2013)
Ld—Landslide dams; *L*—Lake; *Vf*—Valley floor; *R*—River

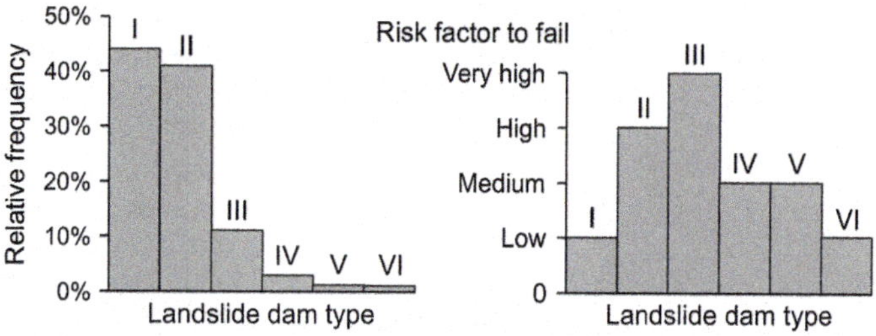

Fig. 2.4 Relative frequency and risk factor to fail for each type of landslide dam (Swanson et al. 1986; Costa and Schuster 1988)

because besides the water they contain, there is also the danger coming from the water accumulated from affluents and adjacent lakes (Swanson et al. 1985). The lakes formed as a result of diluvium barriers of the IVth and Vth categories can only pose a threat when they appear in tight valleys, and the size of the dam is significant (Butler et al. 1986). In the case of lakes of Type VI there is practically no complete blockage, as the river continues to flow over the dam, following the initial orientation of the valley. This is the reason for which the accumulation of water only reaches a low level and, implicitly, the danger of extreme phenomena is low as well (Fig. 2.4).

Based on this classification and having analysed a considerable number of lacustrine accumulations formed by diluvium barriers on our planet, the authors reached the conclusion that most common lakes are the ones belonging to the second cate-

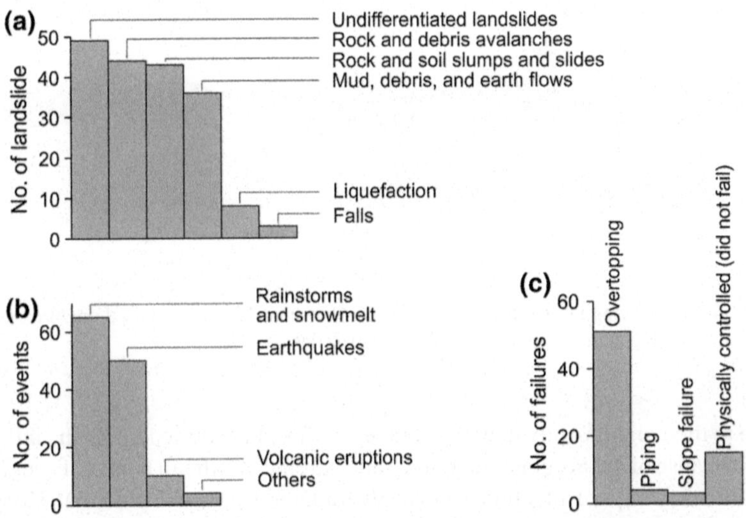

Fig. 2.5 Graphs showing: **a** Distribution of landslide dams by type of landslide, based on 183 cases (Classification based on Varnes 1978); **b** Causes of landslides that have formed dams, based on 128 cases (Costa and Schuster 1988); **c** Modes of failure of landslide dams, based on 55 failures from the literature and the authors' experience (Costa and Schuster 1987, 1988)

gory (44%), followed by the ones in the third category (41%) and the first category (11%). The lakes formed by dams of the sixth category (3%) and the ones in the fourth and fifth categories (<1%) are the least common (Swanson et al. 1986; Costa and Schuster 1988).

A category which is often seen among natural dam lakes include the lakes formed behind the barriers of ice and snow, or behind the debris caused by the movement of glaciers. Most frequently, they appear in areas affected by an excessive continental climate, in glacial or periglacial areas, and some are formed by the movement of masses of snow in high altitude mountainous areas. Also included in this category are the moraine-dammed lakes and the lakes formed behind the mixture of rocks, soil and vegetation amassed by avalanches in valleys (Ballantyne and McCann 1980; Costa and Schuster 1988; Rudoy 2002a, b; Benn et al. 2006; Romanescu et al. 2013) (Fig. 2.5).

Most of the time, these lakes are temporary and cause floods upstream after the disintegration of the dam as well, besides flood waves downstream. Icelanders call this phenomenon "jökulhlaup"—flood caused by the sudden release of water accumulated behind ice (Thorarinsson 1953). In Romania, along the valley of the river Bistriţa, the locals use the term "zăpor" for the ice dams which hinder the flow of the river for a short period, causing water accumulations which can sometimes become dangerous for the population living near the sector of the river affected by the blockage.

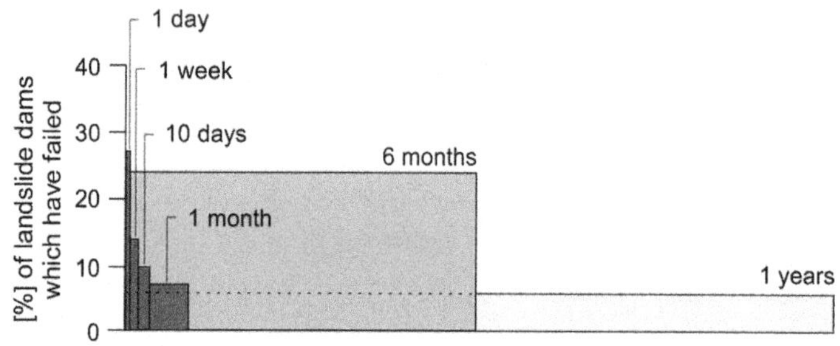

Fig. 2.6 Graph showing length of time before failure of landslide dams, based on 73 cases from the literature and the authors' experience (Costa and Schuster 1988)

Due to their significantly dynamic character, the lakes formed under the influence of, and because of the accumulation of ice are extremely difficult to classify, and most of the time they require attention exclusively for their destructive potential. However, one can still find a series of descriptions of lacustrine basins in the specialty literature, one of the most intuitive of them belonging to Blachut and Ballantyne (1976).

Natural dam lakes can last from up to a few minutes to thousands of tens of thousands of years, based on the density and material of the dam, hydrological particularities of the area and last but not least, the pattern of relief forms (Fig. 2.6). The classification of these lacustrine basins is not an easy task and demands an approach which might differ from one researcher to another. The typology presented in this paper represents a mere review of the information retrieved from international literature.

2.3.2 *The Distribution of Natural Dam Lakes on Earth*

Most of the natural dam lakes are formed in mountainous and submountain areas, where the relief is favourable and there is a high predisposition to initiate slope processes. The greatest number of lakes belonging to this category is found in The Andes, The Rocky Mountains, Pamir, The Alps, Tian Shan and others (Huss et al. 2007). Lakes are very frequently formed in volcanic areas, due to the dams created by lava leakage and pyroclastic materials, but also as a consequence of seismic activity. Such lakes are found in China, Japan, Taiwan, Tadjichistan, U.S.A., Mexico and others. The lakes formed behind ice dams are mostly associated with subarctic areas: Alaska, Labrador, Taymyr, Lapland, Tierra del Fuego etc. (Marsh and Neumann 2003; Romanescu 2000) (Fig. 2.7).

Landslides are the events which create most of the natural dam lakes on our planet. In the specialty literature, certain lakes are usually used as illustrative examples: Lower Gros Ventre, formed in 1927, Cerro Condor-Sencca in Peru, formed in 1945, Yashinkul on the river Isfayramsay in the South-Central area of Russia, Cache Creek in Northern California (Alden 1928; Glazzyrin and Reyzvikh 1968; Scott 1985).

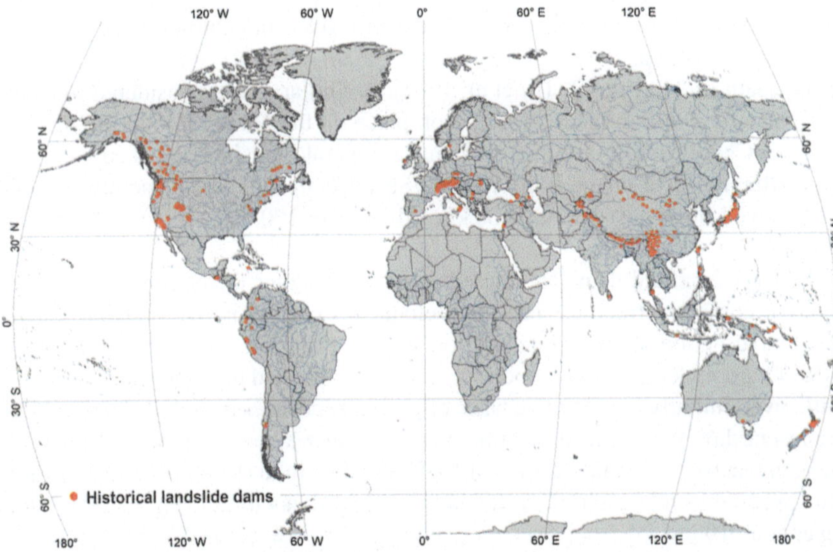

Fig. 2.7 The distribution of 463 historical landslide dams and associated natural reservoirs that have been recorded throughout the world (Costa and Schuster 1988)

In the numerous studies dedicated to this topic there are mentions of lakes which were once of considerable size, but did not maintain their appearance because of the fragility of the dams. Holmes (1965) brought proof of the existence of such lakes in Switzerland, in the year 563, as well as the existence of a water accumulation in the centre of Java, in 1006. It was a lake of the same type that Davis (1882) wrote about: "In Oisans (France), in the Western Alps, a landslide caused, in 1181, the formation of a natural dam lake with a length of about 10 km, known under the name of Saint Laurent Lake, and the farmers in that valley became fishermen overnight" (Davis 1882; Costa and Schuster 1987). Among the ones that have maintained their features up to present-day, being considered the most long-lasting lakes, we mention: Saint André Lake in France, formed in 1248, and Busyû in Japan, formed in 1590 (Jakobson et al. 2007).

In the past, towards the end of 1840 and at the beginning of 1841, there was a landslide which blocked the valley of the Indus, in the Parbat Nanga area in Kashmir (Pakistan). In May 1841, the lake, 64 km in length and 300 m in depth, was drained in just 24 h because of the destruction of the dam. Also, in 1893, along the upper course of the Ganges, there was a huge landslide which eventually led to the formation of a lake with a depth of 250 m. Eight months later, the water overflowed, and 100 million tonnes of water were drained in just two hours, generating a wave with a height of 15 m, propagated along 100 km (Schuster and Wieczorek 2002).

Natural dam lakes are dangerous in some cases, causing floods through the accumulation of a large quantity of water and then through the release of this water over the dam, as flood waves, or through literally breaking the diluvium. The most recent events of this type occurred in 1933, on the Min river, in central China, when the Deixi basin, totalizing 400 million tonnes of water suddenly shed its water, causing the death of 2423 people (Li et al. 1986). A similar event, considered to be the most dramatic one in human history, took place on the Dadu river, in the Sichuan province (China), when, after an earthquake, several landslides blocked the entire valley. After ten days, on the 10th of June 1786, the dam collapsed, generating a wave which moved along the river for 1400 km, killing over 100,000 people (Schuster and Evans 2011).

The largest natural dams are caused by several trigger factors and usually they display a complex structure. The most eloquent example is the landslide caused during the eruption of the St. Helens volcano, in 1980. The dam, 2.8 km in length, made of hard rocks, soil and pyroclastic materials, blocked the valley of North Fork Toutle river, initially forming five lakes of great size, from which three have held till the present day. We can also mention here two other lakes, the Grande Rivière du Chene in Quebec (Canada), formed in 1943, and Yashinku (Russia) formed in 1966, both appeared as the result of earthquakes causing rocks to fall (Pusharenko 1982; Meyer et al. 1986).

The high level of seismic activity was also responsible for the landslides in 1889 on the Totsu river in Japan. The surface that was affected was as large as 1000 km^2, causing the formation of 53 temporary lakes (Swanson et al. 1986). In 1929, in New Zealand (The Southern Island), an earthquake with a magnitude of 7.60 on the Richter scale caused a landslide which led to the formation of 11 lakes (Adams 1981). The greatest number of natural dam lakes appeared almost simultaneously as a result of an earthquake was recorded in 1783, in Calabria (Italy), when as many as 215 were formed (Cotecchia 1978).

Lake Sarez is the deepest natural dam lake on Earth and it was formed due to a landslide caused by an earthquake with a magnitude of 6.5–7 on the Richter scale. This event occurred on the 6th of February 1911, in Central Asia (Tadjikistan), on Murghab river (Pamir Plateau). At present, the basin is 505 m in depth, with a surface of 79.7 km^2, 55.8 km in length, 3.3 km maximum width, a volume of 16.07 km^3, and has an estimated durability of about 300 years (Bolt et al. 1975).

Lake Clear in California is the oldest natural dam lake formed through damming with volcanic lava, being approximately 480,000 years old. It is also the current record holder as far as surface is concerned (180 km^2), even though at present it actually consists of two smaller lakes separated after a new landslide. Taken together, the two lakes have a length of 31 km, a maximum width of 13 km and an average depth of 8.2 m (Romanescu et al. 2013).

Regarding the natural dam lakes formed due to the activity of glaciers, the oldest one in Europe is considered to be Chaillexon (Brenets), in the valley of the river Doubs (France). It was formed during the glacial withdrawals Wurm I and II, 12,000 years ago, through the diluvial dam caused by two landslides which blocked the valley with compacted deposits. At present, 90% of it is clogged up, but through paleolimnological studies it has been established that initially it had a length of 15 km, a volume of 90 mil. m^3 and a surface of 750 ha (Bichet et al. 1995; Di-Giovanni et al. 1998).

Even if lacustrine basins formed as a result of glacial valleys being blocked are usually reduced in size, mention can be made of a series of lakes with rather large surfaces. The most significant examples are the ones in the upper catchment area of the Indus, formed behind the Chong Khumdan glacier, or the ones in the proximity of the Shyok glacier on the affluent bearing the same name, which caused intense flooding in 1929 (Mason 1929; Gunn 1930; Hewitt 1982; Costa and Schuster 1987).

The most important glacial dams are represented by the moraine accumulations formed in high altitude areas. They are the most dangerous, as well, since they cause significant material damage and loss of human lives when they are destroyed. Among the best known occurrences of this type, we can mention here the ones in Peru, Nepal, India, Pakistan, Rusia, Canada, Austria and Argentina (Yesenov and Degovets 1979; Burgusser et al. 1982; Eisbacher and Clague 1984; Clague et al. 1985; Vuichard and Zimmermann 1987) (Table 2.3).

2.4 The Research of Natural Dam Lakes in Romania

In Romanian literature, natural dam lakes are mentioned in the classification of lacustrine formations in a simplified version, compared to the diversity displayed by international literature. The main reason for which this type of lake is absent from national studies—lakes formed through water being impounded behind ice dams, formed by avalanches or the debris and materials they carry along, the ones formed by lava discharges, the ones formed through biological activity a.o.—is most probably their low incidence. The category which is the most common on the Romanian territory is the one consisting of natural dam lakes formed through caving-in and gravitational landslides (Gâştescu 1963, 1979, 1998; Romanescu et al. 2013).

In our country, the most favourable conditions for the formation of natural dam lakes are met in the Eastern Carpathians, the Subcarpathians of Curvature and the Moldavian Subcarpathians, as well as in some isolated areas in the Moldavian Plateau. The highest susceptibility is found in the Carpathian flysch, where the alternation of permeable and impermeable layers, the lithological alternation of soft and hard rocks and the pronounced declivity of the relief are the cause of slope processes, and, in some cases, to the formation of lacustrine reservoirs.

The first mentions in Romanian literature include the natural dam lake Betiş, formed due to a landslide which occurred in 1956, in the area of the Maramureş Mountains. The diluvial mass blocked the Ampol source, an affluent of the river Vaser, forming an accumulation totalizing 2 ha, with a maximum depth of 9 m. The lake did not last for long, becoming completely drained in just two years. Another example is Lake Bolătău, on the river Zăbala, an affluent of the river Uz, formed in 1983. Its initial configuration and depth have changed significantly, at present having a depth of 3.5 m and a surface of 7 ha. In 1977, another natural dam lake was formed along the same source, but only lasted for three weeks (Gâștescu 1979).

Most of the natural dam lakes formed in Romania are found in areas which are difficult to reach, in mountaineous areas, and they usually have a small size and a

Table 2.3 Well-documented historical landslide dams from around the world which did not fail (Costa and Schuster 1991; Romanescu et al. 2013; completed by Mihu-Pintilie et al. 2014)

Date	River or lake	Landslide		Trigger	Dam				Dam materials	Lake		References
		Type	Vol. (m^3)		Type	H (m)	L (m)	W (m)		L (m)	Vol. (m^3)	
Afghanistan–Bamiyan Province												
1960s (early)	Ajar R. Chiltan L.	Fall, rock	–	Earthquake	–	–	–	–	Many rock block	–	–	Costa and Schuster (1991)
Austria–Tirol province (Velber Valley)												
1495	Velber B. Hintersee L.	Fall, rock	–	Earthquake	–	–	–	–	Many rock block	–	–	Montandon (1933)
Chile (southern)												
1960/V/22	Pellaifa L.	Avalanche, rock	–	Earthquake	–	8	–	–	Volcanics	–	–	Weischet (1963)
China–Gansu province (Zhouqu)												
1981/IV/11	Bailong R.	Flow, earth	4,000,0000	–	–	25	–	–	Phyllite, schis, loess	–	19,000,000	Li et al. (1986)
– Qinghai province (Bitang)												
1961/XI/14	Bitang C.	Slump	8,000,0000	–	–	65	–	–	Soft Tertiary clay stone	1000	4,200,000	Li et al. (1986)
Colombia–Boyaca state (Paz del Rio)												
1979/XII/10	Chicamocha R.	Slump	1,420,0000	Rainfall	VI	15	100	600	Claystone, sandstone	3000	–	Costa and Schuster (1991)

(continued)

Table 2.3 (continued)

Date	River or lake	Landslide		Trigger	Dam				Dam materials	Lake		References
		Type	Vol. (m^3)		Type	H (m)	L (m)	W (m)		L (m)	Vol. (m^3)	
France–Drome department												
1442	Drome R.	Avalanche, rock	2,000,000	Earthquake (?)	V	–	600	–	Rock debris, very large blocks	8000	–	Costa and Schuster (1991)
– Haute-Savoie Department (near Servoz)												
1471/II/?	Arve R.	Avalanche, rock	–	–	–	–	–	–	Limestone and shale	–	–	Montandon (1933)
Georgia–Adzhariya Region (near Ambrolauri, NE of Kutaisi)												
1972	Ritseuli R.	–	30,000,000	–	–	20	–	–	Oligocene and Miocene clays	–	–	Costa and Schuster (1991)
Guatemala (15 km SW of Tecpan)												
1976/II/4	Los Chocoyos R.	Avalanche, debris	750,000–1,000,000	Earthquake	III	20–50	300–400	800	Pleistocene pumice	300	–	Harp et al. (1981)
Hungary–Borsod-Abauj-Zemplen County												
1929	Szohony S. Arlo L.	Slide, rotational	5,000,000	Human-caused (mining)	II	90	100	400	Miocene clay	650	150,000	Costa and Schuster (1991)
Italy–Calabria Region												
1783/II/5–6	Lindo R. San Bruno L.	Slide, earth	–	Earthquake	IV	32	80	–	Pliocene sandy marly clays	1380	–	Cotecchia (1978)

(continued)

Table 2.3 (continued)

Date	River or lake	Landslide		Trigger	Dam				Dam materials	Lake		References
		Type	Vol. (m^3)		Type	H (m)	L (m)	W (m)		L (m)	Vol. (m^3)	
1783/II/5–6	Duverso T. Speziale L.	Slide, earth	–	Earthquake	IV	12	100	–	Pliocene sandy, marly clays	470	–	Cotecchia (1978)
1783/II/5–6	Duverso T. Coluce ed Arena L.	Slide, earth	–	Earthquake	II	26	170	–	Pliocene sandy, marly clays	830	–	Costa and Schuster (1991)
– Lombardia Region (Val Bregaglia, Piuro)												
1618/IX/4	Mera R.	Avalanche, rock and debris	3,000,000–4,000,000	Rainfall	–	–	–	–	Gneiss and amphibolite debris	–	–	Eisbacher and Clague (1984)
(Val Codera)												
1988/VII/14	Codera S.	Flow, debris	–	Rainfall	–	3	–	–	–	200	–	Costa and Schuster (1991)
(Valtellina)												
1987/VII/28	Adda R. Val Pola L.	Avalanche, debris	35,000,000	Rainfall and undercutting	III	33	2700	550	Gabbro, diorite, gneiss rocks, debris	2900	22,000,000	Costa and Schuster (1991)

(continued)

Table 2.3 (continued)

Date	River or lake	Landslide		Trigger	Dam				Dam materials	Lake		References
		Type	Vol. (m^3)		Type	H (m)	L (m)	W (m)		L (m)	Vol. (m^3)	
(Italy continued)–Piemonte Region (Antronapiana)												
1642/VII/27	Ovesca T. Antronapiana L.	Avalanche, rock	12,000,000	–	–	45	650	1500	Blocks of gnefss	800	–	Montandon (1933); Eisbacher and Clague (1984)
– Trentino-Aldo Adige Region (Molveno)												
200–1000 A.D.	Sarca R. tributary	Avalanche, rock	375,000,000	–	–	130	1500	3250	Limestone and dolomite	4500	–	Costa and Schuster (1991)
Veneto Region (Cordevole Valley, Lago di Alleghe)												
1771/I/11	Cordevole T. L. of Alleghe	Slide, rock/avalanche, rock	20,000,000	–	II	80	500	1750	Marl, limestone debris	4000	–	Eisbacher and Clague (1984)
Japan–Fukushima Prefecture (Iwashiro Province)												
1618/IX/4	Mera R.	Avalanche, rock and debris	3000,000–4,000,000	Rainfall	–	–	–	–	Gneiss and amphibolite	–	–	Swanson et al. (1986)
– Gifu Prefecture (Tokuyama Village)												
1965/IX/?	Ibi R. tributary	Glide, block	1800,000	Rainfall	II	25	200	250	–	650	1300,000	Costa and Schuster (1991)

(continued)

Table 2.3 (continued)

Date	River or lake	Landslide		Trigger	Dam				Dam materials	Lake		References
		Type	Vol. (m^3)		Type	H (m)	L (m)	W (m)		L (m)	Vol. (m^3)	
– Kanagawa Prefecture (Hatano City)												
1923/IX/1	Shinsei L.	Slump, earth	230,000	Earthquake	II	10	100	200	Sand, gravel, loam	200	37,000	Costa and Schuster (1991)
– Nagano Prefecture (Azumi Village)												
1915/VI/6	Azusa R.	Flow, debris	1700,000	Volcanic eruption	–	4.5	–	–	Andesite	2000	530,000	Costa and Schuster (1991)
1926/VII/23	Azusa R.	Flow, debris	–	Volcanic eruption	–	10	–	330	Andesite	2300	1200,000	Costa and Schuster (1991)
(Hakuba Village)												
1891/VI/?	Matsu R.	Slide, rock/avalanche, earth	3200,000	Rainfall	II	55	500	230	Sedimentary rocks	500	3100,000	Costa and Schuster (1991)
(Japan continued)–(Nagano Prefecture, Mount Ontake)												
1984/IX/14	Otaki R.	Avalanche, debris	36,000,000	Earthquake and rainfall	III	40	250	2500	Volcanic debris	1000	–	Costa and Schuster (1991)
(Otari vilage)												
1971/VII/16	Hime R.	Slide, earth	300,000	Rainfall	I	–	80	80	Altered tuff, mudstone, sandstone	–	–	Costa and Schuster (1991)
– Nara prefecture (Nishitotsugawa village)												
1889/VIII/?	Totsu R. tributary	Slide, rock and earth	4,400,000	Rainfall	II	25	130	250	Sedimentary rocks	250	110,000	Costa and Schuster (1991)

(continued)

Table 2.3 (continued)

Date	River or lake	Landslide		Trigger	Dam				Dam materials	Lake		References
		Type	Vol. (m^3)		Type	H (m)	L (m)	W (m)		L (m)	Vol. (m^3)	
(Nishiyoshin village)												
1982/VIII/4	Niu R.	Slump, rock/slide, debris	610,000	Rainfall	II	15	50	180	Slate	3000	1,300,000	Swanson et al. (1986)
(Oji Town)												
1931–1932	Yamato R.	Slide, rock and debris	22,000,000	Rainfall	VI	20–36	50	170	Sandstone, tuff, clay	–	10,000,000	Swanson et al. (1986)
– Shizuoka Prefecture (Izu Peninsula, Nanamawari)												
1978/I/14	Tajiri R.	Slide, debris	50,000	Earthquake	II	10	20	125	Volcanic debris	–	–	Chigira and Masahiro (1989)
(Shizuoka city)												
1702	Abe R.	–	120,000,000	Earthquake	III	30	500	650	Sandstone and slate	–	4,700,000	Costa and Schuster (1991)
– Tochigi Prefecture (Nikko City)												
1949/XII/26	Oya R.	Slide, rock and earth	9,000	Earthquake	II	10	50	100	Scoria	–	3300	Costa and Schuster (1991)
– Wakayama Prefecture (Hanazono village)												
1953/VII/18	Arida R. tributary	Slide, rock	250,000	Rainfall	II	25	100	100	Sandstone and slate	200	60,000	Costa and Schuster (1991)

(continued)

Table 2.3 (continued)

Date	River or lake	Landslide		Trigger	Dam				Dam materials	Lake		References
		Type	Vol. (m^3)		Type	H (m)	L (m)	W (m)		L (m)	Vol. (m^3)	
(Japan continued)–Yamagata Prefecture (Ooe Town)												
1984 4/11	Kose R.	Slump	5600,000	Snowmelt	II	20	–	90	Mudstone, siltstone, tuff	500	–	Costa and Schuster (1991)
Kazachstan (near Alma Alta)												
1966–1967	Malaya Alma-Atinka R.	–	2,500,000	Human-caused (explosions)	II	110	–	–	–	–	6200,000	Yesenov and Degovets (1979)
Kyrgyzstan–Tien Shan Mts.												
1989/VI/11	Utch-Terek R. (trib. of Naryn R.)	–	3000,000	Human-caused (explosions)	II	42	–	–	–	–	–	Costa and Schuster (1991)
Macedonia–Gradot Hill												
1956/IX/5	Vatasha R.	Slump, rock and soil	20,000,000	Undercutting	II	70	400	800	Tuff, conglomerate, clay	–	–	Costa and Schuster (1991)
– Jovatz River Valley												
1977/II/15	Jovatz R.	Slide, debris	2,00,000,000	Rainfall	II	15–20	150	700	Claystone, sandstone, debris	2500	500,000	Costa and Schuster (1991)

(continued)

Table 2.3 (continued)

Date	River or lake	Landslide		Trigger	Dam				Dam materials	Lake		References
		Type	Vol. (m^3)		Type	H (m)	L (m)	W (m)		L (m)	Vol. (m^3)	
New Zealand–Hawke's Bay Area (North Island)												
1968/VII/1	Tuki Tuki R. Folger L.	Slide, rock/avalanche, debris	>7,65,000	Undercutting	II	15	–	–	Sandstone and argillite debris	–	–	Adams (1981)
(Near Kairak)												
1976/X/24	Ponui S.	Slide, block and debris	2,500,000	Rainfall	II	50–60	?	570	Sandstone	–	2890,000	Pettinga (1987)
– Nelson Area (South Island)												
1929/VI/17	Matiri R. Lower L.	Slide, rock	–	Buller earthquake	–	–	–	–	Tertiary sandstone	–	–	Adams (1981)
(New Zealand continued)–(Nelson Area, South Island)												
1929/VI/17	Matiri R. Upper L.	Slide, rock	–	Buller earthquake	–	–	–	–	Tertiary sandstone	–	–	Adams (1981)
– Wellington Area (North Island)												
1855/I/23	Ruamahanga R.; Bruce L.	Slide, rock	–	Earthquake	II	–	–	–	Sedimentary rocks	–	–	Adams (1981)

(continued)

Table 2.3 (continued)

Date	River or lake	Landslide		Trigger	Dam				Dam materials	Lake		References
		Type	Vol. (m^3)		Type	H (m)	L (m)	W (m)		L (m)	Vol. (m^3)	
Norway–Ulvadal Valley												
1960/VI/26	Ulvadal R. Ulvadalsvatn L.	Slide, debris	–	Rainfall	II	2–3	–	–	Boulders, birch trees, and earth	–	–	Costa and Schuster (1991)
Pakistan–Kashmir (Hunza)												
1976	Hunza R.	Flow, debris	–	–	–	20	–	–	–	12,000	–	Costa and Schuster (1991)
Papua New Guinea–East New Britain Province												
1985/V/11	Bairaman R. (NE)	Slide, rock/avalanche	150,000,000	Earthquake	III	50	500	1000	–	3500	2100,000	King and Loveday (1985)
– East Sepik province (near Maprik)												
1935/IX/20	Clearwater C. Imbia L.	Slide, rock and earth	2,000,000	Earthquake	III	45	100	200	Conglomerate and sandstone	500	1000,000	King (1986)
– West New Britain province												
1985/V/11	Tiaru R.	Slide, rock and debris	5,000,000	Earthquake	–	50	200	500	–	1000	2000,000	King et al. (1989)

(continued)

Table 2.3 (continued)

Date	River or lake	Landslide		Trigger	Dam				Dam materials	Lake		References
		Type	Vol. (m^3)		Type	H (m)	L (m)	W (m)		L (m)	Vol. (m^3)	
Peru–Ancash Department												
1970/V/31	Nepena R.	Slide, rock	–	Earthquake	–	20	?	–	Fractured granodiorite blocks	480	–	Plafker et al. (1971)
1970/V/31	Shacsha R. Llanganuco L.	Fall, rock/avalanche, debris	500,000	Earthquake	II	–	–	–	–	–	–	Ericksen et al. (1970)
(Peru continued)–(Ancash Department, Recuay)												
1970/V/31	Santa R.	Slump	25,000,000	Earthquake	VI	–	150	300	Fluvio-glacial sediments	700	–	Plafker et al. (1971), Ericksen et al. (1970)
Philippines–Luzon Island, Benguet province (Iriga Mts.)												
1628 (?)	Naporoc R. tributary Buhi L.	Avalanche, debris	1,500,000,000	–	III	90	–	–	Volcanic debris	5000	–	Aguila et al. (1986)
Romania–Carpathian Mts. (Stânişoarei Mts.)												

(continued)

Table 2.3 (continued)

Date	River or lake	Landslide		Trigger	Dam				Dam materials	Lake		References
		Type	Vol. (m^3)		Type	H (m)	L (m)	W (m)		L (m)	Vol. (m^3)	
1991/VI–VIII/?	Cuejdel C. Cuejdel L.	Slide, rock	–	Rainfall, earthquake (?)	II	25	1050	1200	Limestone and flysch rock	1100	925,000	Mihu-Pintilie et al. (2014)
1991/VI–VIII/?	Cuejdel C. Constelației L.	Slide, rock	–	Rainfall, earthquake (?)	II	5–10	800	900	Limestone and flysch rock	50	–	Mihu-Pintilie et al. (2014)
1991/VI–VIII/?	Cuejdel C. Albastru L.	Slide, rock	–	Rainfall, earthquake (?)	II	5	300	500	Limestone and flysch rock	20	–	Mihu-Pintilie et al. (2014)
(Hășmaș Mts., Ucigașul Peak)												
1837/VII/?	Bicaz R. Roșu L.	Slide, rock	–	Rainfall, earthquake (?)	II	–	–	–	Limestone and flysch rock	1350	721,000	Romanescu et al. (2013)
Serbia–Zavoj (near Temska)												
1963/II/23–27	Visotchitza R.	Slide, debris	4,000,000	Snowmelt and rainfall	II	35	150	500	Clay with sandstone	5000	14,000,000	Costa and Schuster (1991)
Spain–Granada (Los Olivares)												

(continued)

Table 2.3 (continued)

Date	River or lake	Landslide		Trigger	Dam				Dam materials	Lake		References
		Type	Vol. (m^3)		Type	H (m)	L (m)	W (m)		L (m)	Vol. (m^3)	
1986/IV/?	Velillos R.	Slump/flow, earth	500,000	Human-caused (mining)	II	3	15	35	Shale and sandstone	–	–	Costa and Schuster (1991)
Switzerland–Berne Canton (Moutier-Court Gorge)												
1937 III/?–IV/?	Birse R. (upper dam)	Slide, debris	2,000,000	Rainfall	V	6	25	80	Limestone, marl, and shale debris	60	?	Costa and Schuster (1991)
1937 III/?–IV/?	Birse R. (upper dam)	Slide, debris	2,000,000	Rainfall	V	9	30	110	Limestone, marl, and shale debris	110	?	Costa and Schuster (1991)
– Valais Canton (Diablerets)												
1749/VII/23	Derborence T. Derborence L.	Avalanche, rock	30,000,000	–	–	–	–	–	Slate and limestone	–	–	Eisbacher and Clague (1984)
Tajikistan–Usoy												

(continued)

Table 2.3 (continued)

Date	River or lake	Landslide		Trigger	Dam				Dam materials	Lake		References
		Type	Vol. (m^3)		Type	H (m)	L (m)	W (m)		L (m)	Vol. (m^3)	
1964/IV/?	Zeravshan R.	–	15,000,000–20,000,000	Earthquake	–	200	400	1800	Granite, earth and rock	–	–	Costa and Schuster (1991)
1911/II/18	Murgab R. Sarez L	Slide, rock	2,000,000,000	Earthquake	II	800	1000	1000	Blocks of sandstone, shale, and carbonaceous rocks	53,000	16,000,000,000	Gasiev (1984)
Turkey–Trabzon province (Catak)												
1988/VI/23	Degirmen R.	Avalanche, rock	250,000	Rainfall	–	6	50	130	Marls, mudstone, limestone, tuff, basalt	300	–	Costa and Schuster (1991)
United States of America–Alaska (near Seward)												
1964/III/27	Box Canyon C.	Slide, rock and debris	–	Earthquake	II	3–5	–	–	Boulders, cobbles, fine material	–	–	Lemke (1967)
(United States of America continued)–(Alaska, near Seward)												
1964/III/27	Jap C.	Slide, rock and debris	–	Earthquake	II	5–8	–	125	Glacial drift	–	–	Lemke (1967)
– California (Humboldt County)												
1983/IV/3	Mattole R. "Marijuana" L.	Slide/flow, debris	9,500,000	Rainfall	III	12	90	400	Large sandstone and conglomerate	2200	–	Costa and Schuster (1991)

(continued)

Table 2.3 (continued)

Date	River or lake	Landslide		Trigger	Dam				Dam materials	Lake		References
		Type	Vol. (m^3)		Type	H (m)	L (m)	W (m)		L (m)	Vol. (m^3)	
(Santa Cruz County)												
1989/X/17	Corralitos C.	Fall, rock	3500	Earthquake	II	–	10–15	20	Sandstone fragments and blocks	200	6000	Costa and Schuster (1991)
1989/X/17	Hinckley C.	Fall, rock	1600	Earthquake	II	5	15	15	Sandstone fragments and blocks, redwood trees	130	–	Costa and Schuster (1991)
1989/X/17	Corralitos C.	Fall, rock	3500	Earthquake	II	3	25	15	Sandstone fragments and blocks	30	1200	Costa and Schuster (1991)
– Colorado (Grand County)												
1985/IV/16	Fraser R.	Slump/flow, debris	3000	Snowmelt	III	1–2	–	50–70	Colluviun from igneous bedrock	300	–	Schuster (1985)
(Gumison County)												
1986/IV/7	Muddy C.	Slide, earth	180,000,000	Snowmelt	VI	–	30	1000	Tertiary mudstone	–	–	Costa and Schuster (1991)

(continued)

Table 2.3 (continued)

Date	River or lake	Landslide		Trigger	Dam				Dam materials	Lake		References
		Type	Vol. (m^3)		Type	H (m)	L (m)	W (m)		L (m)	Vol. (m^3)	
– Montana (Glacier County, Glacier National Park)												
1910	Otatso C. Slide Pond	Avalanche, rock	7,000,000	Rainfall	V	5	100	120	Limestone blocks and fragments	400	–	Butler et al. (1986)
(United States of America continued)–(Montana, Glacier County, Glacier National Park)												
1910	Otatso C. Slide L.	Avalanche, rock	7,000,000	Rainfall	V	5	100	120	Limestone blocks and fragments	1000	–	Butler et al. (1986)
(Madison County)												
1959/V/17	Madison R. Earthquake L.	Slide, rock	21,000,000	Earthquake	II	60–70	500	1600	Fragments of gneiss, schist, and dolomite	10,000	–	Hadley (1964), Harrison (1974)
– Oregon (Baker County)												
1984/VIII/?	Powder R.	Slide, rock/avalanche, debris	6,000,000	Rainfall	II	9	60	200	Basalt debris	–	290,000	Costa and Schuster (1991)
(Tillamook County)												
1964/XII/?	Wilson R.	Flow, debris	450,000	Rainfall	–	20	–	135	Mud, trees, rock debris	–	–	Costa and Schuster (1991)

(continued)

Table 2.3 (continued)

Date	River or lake	Landslide		Trigger	Dam				Dam materials	Lake		References
		Type	Vol. (m^3)		Type	H (m)	L (m)	W (m)		L (m)	Vol. (m^3)	
1991/IV/4	Wilson R.	Slide, debris	760,000	Rainfall	II	6	50	100	Volcanic rock colluvium	490	–	Costa and Schuster (1991)
– Utah (Kane County, Hackberry Canyon)												
1988/IV/?	Hackberry C.	Fall, rock	–	Undercutting	II	10	30	–	Large sandstone blocks	75	–	Costa and Schuster (1991)
1984/V/13	Cottonwood C.	Flow, debris	–	Snowmelt	III	2.5	20	50	Silt and large limestone	50	–	Costa and Schuster (1991)
(Utah County, Provo Canyon, 16 km NE of Provo)												
1938/VII/13	Provo R.	Flow, debris	–	Rainfall	–	–	–	100	–	–	–	Woolley (1946)
(Utah County, Thistle)												
1983/IV/?	Spanish Fork R.	Slide, debris	22,000,000	Snowmelt	II	63	200	450	Sedimentary rocks	6000	78,000,000	Costa and Schuster (1991)
(United States of America continued)–Washington (Cowlitz county)												
1980/V/18	South Fork Castle C. Castle L.	Avalanche, debris	2,800,000,000	Volcanic eruption	III	37	–	–	Volcanic debris	2000	24,000,000	Meyer et al. (1986)
1980/V/18	Coldwater C. Coldwater L.	Avalanche, debris	2,800,000,000	Volcanic eruption	III	71	–	–	Volcanic debris	3500	83,000,000	Meyer et al. (1986)

(continued)

Table 2.3 (continued)

Date	River or lake	Landslide		Trigger	Dam				Dam materials	Lake		References
		Type	Vol. (m^3)		Type	H (m)	L (m)	W (m)		L (m)	Vol. (m^3)	
(Skamania County)												
1980	North Fork Toutle R. Spirit L.	Avalanche, debris	2,800,000,000	Volcanic eruption	III	69	–	–	Volcanic debris from Mount St. Helens	5000	330,000,000	Schuster (1985), Meyer et al. (1986)
(Thurston County)												
1990/IX/16	Nisqually R.	Slump	200,000	Undercutting (?)	I	–	240	170	Tertiary fine-grained volcano-clastic and till	1000	–	Costa and Schuster (1991)
(Whatcom County)												
1983/I/10	Whatcom tributary L.	Slide, rock	–	Rainfall	I	10	–	250	Sandstone	–	–	Costa and Schuster (1991)
– Wyoming (Teton County)												
1909–1910	Gros Ventre R.; Upper Slide L.	Flow, earth	–	Snowmelt	III	–	1000	3000–4000	Sandstone and limestone	2000	–	Keefler and Love (1956)
(Grand Teton National Park)												
1941/VIII/9–10	Leigh C.	Flow, debris	–	Rainfall	III	–	100	–	Igneous rock debris	300	–	Fryxell and Horberg (1943)

short lifespan. Here, we can mention the Black Lake, on an affluent of the river Buzău (surface—1.6 ha, depth—5 m), the Green Lake on the river Şuşiţa (surface—0.5 ha, depth—4 m), Tăul Zânelor, near Lake Colibiţa (surface—0.3 ha, depth—4 m), Lake Izvorul Măgurii on the river Ilva (surface—0.2 ha, depth—2 m), Devil's Lake formed through the caving in of a cave wall in Nera's Gorge, Lake Veselaru near Roznov (Piatra Neamţ) a.o.

Iezerul Sadovei Lake is considered to be the Romanian natural dam lake with the longest lifespan; it is located in the Feredeu Mountains (Eastern Carpathians), formed after massive gravitational landslides which dammed an affluent of the Sadova source (a left tributary of Moldova river) more than 500 years ago. This age is certified by a medieval document issued on the 12th of July 1594 (Mîndrescu et al. 2010). At present, the lake is extremely clogged, a large quantity of the initial water being drained after a breach in the natural dam, but also as a result of faulty management. In the same area, a few kilometres away, a second dam lake was formed, on the Bolătău source, apparently having a considerable age as well. Unlike the former, the latter displays a less significant degree of clogging and a better resistance throughout time, being also smaller (Mîndrescu et al. 2013).

The Red Lake is perhaps the best known natural dam lake in the Eastern Carpathians; it was formed in 1837, as a result of a landslide starting from under Ucigaşu Peak–Hăghimaş Mountain, the diluvial mass filling the valley of the river Bicaz. From its formation up to the present, the lake has been the focus point for a record number of limnological studies regarding natural dam lakes in Romania (Bojoi 1968; Cărăuşu et al. 1971; Ghenciu 1968; Ghenciu and Cărăuşu 1970; Mihăilescu 1940; Pandi and Magyari 2004; Pandi and Buzila 2004; Pişotă 1956; Romanescu 2009; Romanescu and Stoleriu 2010; Romanescu et al. 2013).

The Red Lake displays a series of characteristics which are typical for the formation of a natural dam lake in the area of flysch mountains, a fact that has always attracted researchers from various fields. Also, the easy access, its lasting in time and the particularities of the relief have transformed it into one of the most important tourist attractions in the East-Carpathians. In 2009, the reservoir was defined by the following values: surface 12.01 ha, length 1.36 km, width 160 m, maximum depth 10.5 m, parameters which closely resemble the information published by the other authors in the past. All these point towards a high level of durability, which helped it reach an age not characteristic for the natural dam lakes formed in our country. Until recently it had been considered the largest natural dam lake in Romania, surface-wise as well as volumetrically, but after 1991, when the natural dam lake Cuejdel was formed (Stânişoarei Mountains), moved to the second position in this classification (Mihu-Pintilie and Romanescu 2011; Romanescu and Stoleriu 2010; Romanescu et al. 2013). The work "Limnology of the Red Lake, Romania. An interdisciplinary study", authored by Romanescu et al. (2013) and published by Springer, basically acted as a model for the present study, which is meant as a continuation of the research on the natural dam lakes formed in the Romanian Carpathians.

Cuejdel Lake has been rarely mentioned in specialty literature, among the first studies which did so being the ones authored by the specialists in Research Centre Pângăraţi (Neamț County). At the beginning of the 80s, all that could be found

Fig. 2.8 Photographs-document of some natural dam lake formed in Eastern Carpathians: **a** Betiş Lake–Maramureş Mts., **b** Iezer Lake–Obcina Feredeului and **c** Red Lake–Hăghimaş Mts.

about this lake consisted in a series of geological and geomorphological analyses that focused on the area of the valley flooded by the lake. Among these, an important one is "Stânişoarei Mountains, geomorphological study", by Ichim (1979), where the author basically describes, for the first time, the landslides that had the tendency of constructing a dam in the valley of the Cuejdel spring.

The 90s witness the first studies exclusively focused on the dam lake formed in 1991, authored by the same researchers in the Pângăraţi Centre. Among these, there is one study which bears a particular importance, since it contained the first bathymetric map of the Cuejdel Lake, made in 1992 and published by Ichim et al. (1996). The studies done by the geographers Ichim et al. (1996), Rădoane (2002) or Rusu et al. (2002) all belong to the same time frame, with all of them using the same bathymetric basis. Subsequently, Cuejdel Lake caught the attention of researchers at the Faculty of Geography and Geology, "Alexandru Ioan Cuza" University of Iaşi (Romania) (Fig. 2.8).

Following these studies, even if the bathymetric data remained unaltered, new information surfaced, regarding the physical, geographical and morphological context of Cuejdel Lake and its surroundings. To that effect, we mention the case-study elaborated by the magazine called "The Hydrology of Continental Environments. Practical Applications" by Minea and Romanescu (2007). However, it was not until 2012 that the first detailed bathymetric mapping of Cuejdel Lake was elaborated (Mihu-Pintilie et al. 2012).

Table 2.4 Well-documented historical landslide dams and morphological parameters of the main natural dam lakes formed on the Eastern Carpathians territory (Gâştescu 1979; Mîndrescu et al. 2013; Romanescu et al. 2013; Mihu-Pintilie et al. 2016)

Lake name	River	Relief unit	Date of landslide dam	Date of dam failure	Lake			References
					Surface (ha)	Volume (m^3)	H max. (m)	
Balătău	Trotuş	Ciuc Mts.	–	–	4.50	–	3.0	Gâştescu (1979)
Bălătău-Ponoare	Cuejdel	Stânişoarei Mts.	–	Did not fail	0.25	5900	7.9	Mihu-Pintilie et al. (2014)
Betiş	Ampol	Maramureş Mts.	1956/?	1958/?	2.00	–	9.0	Gâştescu (1979)
Black L.	Buzău	Buzău Mts.	–	–	1.6	–	5	Gâştescu (1979)
Blue L.	Cuejdel	Stânişoarei Mts.	1991/VI–VIII/?	Did not fail	0.1	–	4	Mihu-Pintilie et al. (2014)
Bolătău	Sadova	Obcina Feredeului	–	Did not fail	0.25	–	5.2	Mîndrescu et al. (2013)
Bolătău	Zăbala	Curvature Carp.	1977/?	1977/?	–	–	–	Gâştescu (1979)
Bolătău	Zăbala	Curvature Carp.	1983/?	Partial failure	7.00	–	3.5	Gâştescu (1979)
Cuejdel	Cuejdel	Stânişoarei Mts.	1991/VI–VIII/?	Did not fail	13.95	925,347	16.5	Mihu-Pintilie et al. (2016)
Constellation L.	Cuejdel	Stânişoarei Mts.	1991/VI–VIII/?	Did not fail	0.34	6541	4.3	Mihu-Pintilie et al. (2014)
Devil's L.	Nera	Nera Mts.	–	Did not fail	–	–	9.3	Romanescu et al. (2013)
Green L.	Şuşiţa	Curvature Subcarp.	–	–	0.5	–	4	Gâştescu (1979)
Iezer	Sadova	Obcina Feredeului	–	Partial failure	0.75	31,414	4.3	Mîndrescu et al. (2013)
Izvorul Măgurii	Ilva	Bârgău Mts.	–	Did not fail	0.2	–	2.0	Gâştescu (1979)
Mocearu	Buzău	Buzău Mts.	–	Partial failure	7.00	–	8.0	Gâştescu (1979)
Red L.	Bicaz	Hăşmaş Mts.	1837/VII/?	Did not fail	12.1	721,404	10.5	Romanescu et al. (2013)
Tăul Zânelor	Colibiţa	Bârgău Mts.	–	Did not fail	0.3	–	4.0	Gâştescu (1979)
Veselaru	Bistriţa tributary	Moldavian Subcarp.	–	–	0.5	–	5	Mihu-Pintilie et al. (2014)

Making a synthesis of the information found in several monographs and limnological studies published in Romanian specialty literature, and also using a series of data provided by the National Administration of Romanian Waters, we have identified a series of morphological parameters of the main natural dam lakes formed on the territory of Romania in the XXth century (Gâştescu 1963, 1979, 1998; Morariu and Savu 1960; Ujvari 1959; Romanescu 2009; Romanescu et al. 2013; Ichim et al. 1996; Rădoane 2002; Pandi and Buzila 2004; Mîndrescu et al. 2010, 2013; Mihu-Pintilie and Romanescu 2011) (Table 2.4).

References

Adams J (1981) Earthquake-dammed lakes in New Zealand. J Geol 9:215–219
Aguila L, Newhall C, Miller C, Listanco E (1986) Reconnaissance geology of a large debris avalanche from Iriga volcano, Philippines. Philippine J. Volcanology 3:54–72
Alden WC (1928) Landslide and flood at Gros Ventre-Wyoming. Am Inst Min Metall Eng Trans 76:347–358
Antipa GR (1910) Regiunea inundabilă a Dunării-Starea ei actuală şi mijloacele de a o pune în valoare. Institutul de Arte Grafice Carol Göbl, Bucharest
Antipa GR (1916) Pescăria şi pescuitul în România. Romanian Academy-National Museum of Natural History, Bucharest
Antonescu CS (1932) Consideraţiuni asupra scăderii pescăriilor din Brateş, Rev Sci Vasile Adamachi 1(2)
Antonescu CS (1967) Biologia apelor. Didactical and Pedagogical Publishing House, Bucharest
Ballantyne CK, McCann SB (1980) Short-lived damming of a high Artic ice-marginal stream, Ellesmere Island, NWT, Canada. J Glaciol 25:487–491
Băcescu M (1942) Observaţiuni biologice şi faunistice în legătură cu Balta Greaca. Rev Sci Vasile Adamachi 2(3)
Benn DI, Owen LA, Finkel RC, Clemmens S (2006) Pleistocene lake outburst floods and fan formation along the eastern Sierra Nevada, California: implications for the interpretation of intermontane lacustrine record. Quat Sci Rev 25:2729–2748
Bichet V, Di-Giovanni C, Richard H, Campy M (1995) Changing patterns of material flow since 12000 BP in the upper Doubs valley (France). Quaternaire 6(2):85–90
Blachut SP, Ballantyne CK (1976) Ice-dammed lakes—a critical review of their nature and behavior, Hamilton, Ontario, vol 6. McMaster University Department of Geography, Discussion Paper, pp 1–90
Blown I, Church M (1985) Catastrophic lake drainage within the Homathko River basin, British Columbia. Canadian Geotechnical J 22(4):551–563
Brătescu C (1915) Lacul Mangalia. In: Bulletin of Romanian Royal Society of Geograph, vol 36. Bucharest
Breier A (1978) Lacurile de pe litoralul românesc al Mării Negre–studiu hidrogeografic. Romanian Academy Publishing House, Bucharest
Bojoi I (1968) Contribuții la sedimentologia Lacului Roșu. Proceedings of the Station for the Biol Geol Geogra Res Stejaru 1:87–105
Bolt BA, Horn WL, Macdonald GA, Scott RF (1975) Geological hazards: earthquakes, tsunamis, volcanoes, avalanches, landslides, floods. Springer, New York
Borcea I (1926) Observations sur le faune des lacs Razelm. Ann Sci Univ Jassy 13(3–4):449–473
Bujor P (1902) Contribution á l'étude de la faune des lacs salés de Roumaine. Ann Sci Univ Jassy 1:149–186
Burgusser HM, Gansser A, Pika J (1982) Late glacial lake sediments of the Indus Valley area, Northwestern Himalaya. Ecol Geol Helvet 75:51–63
Butler DR, Oelfke JG, Oelfke LA (1986) Historic rockfall avalanches, NE Glacier National Park, Montana, USA. Mt Res Dev 6:1–11
Cărăuşu S, Ghenciu V, Timofte LI (1971) Unele date cu privire la trăsăturile hidrofizice ale Lacului Roşu din perioada August 1968–Septembrie 1969. Sci Ann Alexandru Ioan Cuza Univ Iași 17(2):425–445
Chang SC (1984) Tsao-Ling landslide and its effect on a reservoir project. In: Proceedings of 4th international symposium on landslides, vol 1. Toronto, pp 469–473
Clague JJ, Evans SG, Blown IG (1985) A debris flow triggered by the breaching of a moraine-dammed lake, Klattasine Creek, British Columbia. Can J Earth Sci 22:1492–1502
Clark TH (1947) Two Recent landslides in Quebec. Transactions of the Royal Society of Canada 41(3):9–18

Clarke GKC, Waldron DA (1984) Simulation of the August 1979 sudden discharge of glacier-dammed flood lake, British Columbia. Can J Earth Sci 21:502–504
Code JA, Sirhindi S (1986) Engineering implications of impoundment of the Indus River by an earthquake-induced landslide. In: Shuster RL (ed) Landslide dams—processes, risk, and mitigation, vol 3. American Society of Civil Engineers, Geotechnical Special Publication, pp 97–110
Cole GA (1968) Desert Limnology. In: Brown GW Jr (ed) Special topics on the physical and biological aspects of arid regions. Desert Limnology, vol 1. Academic Press, New York, pp 1–635
Cole GA (1994) Textbook of limnology, 4th edn. Waveland, Illinois
Costa JE, Schuster RL (1987) The formation and failure of natural dams. Geol Soc Am Bull, US Geological Survey Report 87-392. Vancouver, Washington
Costa JE, Schuster RL (1988) The formation and failure of natural dams, 2nd edn. Geol Soc Am Bull 100(7):1054–1068
Costa and Schuster (1991) Documented historical landslide dams from around the World. U.S. Geological Survey Open-File Report 91-239, Washington
Cotecchia V (1978) Systematic reconnaissance mapping and registration of slope movements. Bull Int Assoc Eng Geol 17:5–37
Cotton CA (1941) Landscape, as developed by the processes of normal erosion. John Wiley and Sons, New York
Crandell DR, Varnes DJ (1961) Movement of the Slumgullion earthflow near Lake City, Colorado, U.S. Geological Survey Professional Paper 424-B:136–139
Davis WM (1882) On the classification of lake basins. In: Proceedings of the boston society of natural history, vol 21, pp 315–381 (Abst. American Naturalist 16:1028–1029)
Di-Giovanni C, Disnar JR, Bichet V, Campy M, Guillet B (1998) Geochemical characterization of soil organic matter and variability of a postglacial detrital organic supply (Chaillexon Lake, France). Earth Surf Process Land 23(12):1057–1069
Dussart B (1966) Limnologie–L'étude des eaux continentales. Ghautier-Villars, Paris
Eisbacher GH, Clague JJ (1984) Destructive mass movements in high mountains-Hazard and management. Geol Surv Can Pap 84(16):1–230
Ericksen GE, Plafker G, Concha JF (1970) Preliminary report on the geologic events associated with the May 31, 1970, Peru earthquake. U.S. Geological Survey Circular 639, Washington
Evans SG (1984) The 1880 landslide dam on Thompson River near Ashcrof, British Columbia. Current Research of Geol Surv Can Pap 84:655–658
Finch RH (1937) A tree ring calendar for dating volcanic event at Cinder Cone, Lassen National Park, California. American J Sci s5-33(194):140–146
Forél FA (1869) Matériaux pour servir à l'étude de la faune profonde du lac Léman. Bull Soc Vaud Se Nat 16:149–169
Forél FA (1892) Le Léman: monographie limnologique. In: Rouge F (ed) vol 1. Librarie de l'Université, Lausanne, p 539
Forél FA (1895) Le Léman: monographie limnologique. In: Rouge F (ed) vol 2. Librarie de l'Université, Lausanne, p 651
Forél FA (1901) Handbuch der Seenkunde: Allgemeine Limnologie. Engelhorn VJ, Stuttgart, p 249
Forél FA (1904) Le Léman: monographie limnologique. In: Rouge F (ed) vol 2. Librarie de l'Université, Lausanne, p 715
Fryxell FM, Horberg L (1943) Alpine mudflows in Grand Teton National Park, Wyoming. Geol Soc America Bull 54(3):457–472
Gasiev E (1984) Study of the Usoy landslide in Pamir. Landslide 1:511–515
Gâştescu P (1959) Limanele fluviatile dintre Ostrov şi Cernavodă (The fluvial lakes between Ostrov and Cernavodă). Meteo Hidro Gospodărirea Apelor 2:25–32
Gâştescu P (1963) Lacurile din R.P. România–geneză şi regim hidrologic. Romanian Academy Publishing House, Bucharest
Gâştescu P (1971) Lacurile din România–limnologie regională. Romanian Academy Publishing House, Bucharest

Gâştescu P (1979) Lacurile Terrei. Albatros Publishing, Bucharest
Gâştescu P (1998) Limnologie şi Oceanografie. HGA Publishing, Bucharest
Ghenciu IV (1968) Regimul termic al apei Lacului Roșu. In: Communications of the Scientific Session, Alexandru Ioan Cuza University Iași, pp 145–160
Ghenciu IV, Cărăuşu S (1970) Câteva date cu privire la dinamica cantitativă a fitoplanctonului din apa Lacului Roșu în perioada iulie 1967–Decembrie 1968. Sci Ann Alexandru Ioan Cuza Univ Iași 16(1):147–163
Glazzyrin GY, Reyzvikh VN (1968) Compulation of the flow hydrograph for the breach of landslide lakes. Soviet Hydrol 5:592–596
Goldman CR, Horne AJ (1983) Limnology. McGraw Hill Book Company, New York
Gunn JP (1930) The Syook flood, 1929. Himalayan J 2:35–47
Hadley JB (1964) Landslides and related phenomena accompanying the Hebgen Lake earthquake of August 17, 1959. U.S. Geological Survey Professional Paper 435:107–138
Halbfass W (1923) Grundzüge einer vergleichenden Seenkunde (Basic features of a comparative lakes customer). Bronträger, Berlin
Harp EL, Wilson RC,Wieczorek GF (1981) Landslides from the February 4, 1976, Guatemala earthquake. U.S. Geological Survey Professional Paper 1204-A:1–35
Harrison A (1974) Madison Canyon slide mass modification by the U.S. Army Corps of Engineers. In: Voight B, Voight MA (eds), Rock Mechanics – The American Northwest, Third International Congress on Rock Mechanics, International Society for Rock Mechanics, Pennsylvania State University College of Earth Sciences Experiment Station, Special Publication 1:138–143
Hewitt K (1982) Natural dams and outburst floods of the Karakoram Himalaya. In: Glen JW (ed) Hydrological aspects of alpine and high-mountain areas. international Association of Hydrological Science Publication, vol 138, pp 259–269
Holmes A (1965) Principles of physical geology, vol 2. Ronald Press, New York
Hutchinson GE (1957) A treatise on limnology-geography, physics and chemistry, vol 1. Wiley, New York
Hutchinson GE (1967) A treatise on limnology-introduction to lake biology and the limnoplankton, vol 2. Wiley, New York
Hutchinson GE (1975) A treatise on limnology-limnological botany, vol 3. Wiley, New York
Hutchinson GE (1993) A Treatise on Limnology-The Zoobenthos, vol 4. Wiley, New York
Huss M, Bauder A, Werder M, Funk M, Hock R (2007) Glacier-dammed lake outburst events of Gornersee, Switzerland. J Glaciology 53:189–200
Ichim I (1979) Munţii Stânişoarei–studiu geomorfologic. Romanian Academy Publishing House, Bucharest
Ichim I, Rădoane N, Rădoane M (1996) Procese geomorfologice cu intervale de recurenţă mare în arealul munţilor flişului. Exemplificări din Judeţul Neamţ (Geomorphological process with high frequencies in flysch mountains area. Examples from Neamț County). In: Studii şi cercetări—Extras 8, Muzeul de Ştiinţe Naturale Publishing, Piatra-Neamţ, p 15
Imboden C (1976) Eaux vivantes–Initiations la biologie des zones humides. Editions Ligne suisse pour la Protection de la Nature, Bâle
Jakobson M, Björck S, Alm G, Andren T, Lindeberg G, Svensson NO (2007) Reconstructing the Younger Dryas ice dammed lake in the Baltic Basin: Bathymetry, area and volume. Global and Planetary Changes 57:355–370
Kalesnik SV (1956) Bazele geografiei fizice (The bases of physical geography). Editura Ştiinţifică, Bucharest
Keefler WR, Love JD (1956) Landslides along the Gros Ventre River, Teton County, Wyoming. Wyoming Geological Association Annual Field Conference 11th:24–28
King JP (1986) Investigation into the damming of the Bairaman River after the earthquake of 11 May 1985, West New Britain Province. Geological Survey of Papua, New Guinea, File E5080:1–17
King WDVO (1934) The Mendoza River Flood of 10-11 January 1934-Argentina. The Geographical J 84(4):321

King J, Loveday I, Schuster RL (1989) The 1985 Bairaman landslide dam and resulting debris flow, Papua New Guinea. Quarterly J Eng Geol Hydrogeol 22(4):257–270
Lemke RW (1967) Effects of the earthquake of March 27, 1964 at Seward, Alaska. U.S. Geological Survey Professional Paper 542-E:1–43
Lepşi I (1935) Die Seen Rumäniens und ihre Bewirtschaftung. Naturhistorisches Museum, Chişinău
Li T, Schuster RL, Wu J (1986) Landslide dams in South–Central China. In: Shuster RL (ed) Landslide dams—processes, risk, and migitation, vol 3. American Society of Civil Engineers, Geotechnical, Special Publication, pp 146–162
Marr JE (1916) The geology of the Lake District and the scenery as influenced by geological structure. Cambridge University Press, Cambridge
Martonne E (1925) Trait de Géographie physique (Stroke of the physical geography). Tome premier, Paris
Mălăcea I (1956) Lacurile alpine şi viaţa din ele. Revista Natura 6:36–45
Marsh P, Neumann N (2003) Climate and hydrology of a permafrost dammed lake in NW Canada. In: Permafrost, Phillips M, Springman SM, Arenson LU (eds) Swets and Zeitlinger, Lisse, pp 729–734
Mason K (1929) Indus floods and Shyok glacier. Himalayan J 1:10–29
Maxim AI (1929) Contribuţii la explicarea procesului de încălzire a apei lacurilor sărate din Transilvania. Rev Muz Min-Geol Univ Cluj 3(1):121–153
Menshutkin VV, Levchenko VF (2010) Model of development of science by the example of limnology. J Evol Biochem Phisiology 46(6):469–479
Meyer GA, Leidecker ME (1999) Fluvial terraces along the Middle Fork Salmon River, Idaho, and their relation to glaciation, landslide dams, and incision rates. A preliminary analysis and river-mile guide. In: Huges SS, Thackray GD (eds) Guidebook to the geology of Eastern Idaho: Pocatello, Idaho Museum of Natural History, pp 219–235
Meyer GA, Sabol MA, Schuster RL (1986) Landslide dammed lakes at Mt. St. Helens, Washington. In: Schuster RL (ed) Landslide dams—processes, risk, and mitigation, vol 3. American Society of Civil Engineers Geotechnical, Special Publication, Virginia, pp 21–41
Mihăilescu V (1924) Vlăsia şi Mostiştea. In: Bulletin of Romanian Royal Society of Geograph, vol 43. Bucharest
Mihăilescu V (1932) Marile regiuni morfografice ale României. In: Bulletin of Romanian Royal Society of Geograph, vol 50. Bucharest
Mihăilescu V (1940) Cum s-a format Lacul Roşu de la intrarea în Cheile Bicazului. In: Bulletin of Romanian Royal Society of Geograph, vol 59. Bucharest
Mihu-Pintilie A, Romanescu G (2011) Morphometric and morphological suitability of the relief from the Crucii (Cuejdel) Lake basin (Stânişoarei Mountains). In: Pandi G, Moldovan F (eds) Proceedings of the air and water components of the environment, Babeș-Bolyai University, Cluj, vol 1. 18–19 Mar 2011, pp 305–313
Mihu-Pintilie A, Romanescu G, Stoleriu C (2012) Morpho–bathymetric parameters of recess Crucii (Cuejdel) Lake (Stânişoarei Mountains). In: Pandi G, Moldovan F (eds) Proceedings of the air and water components of the environment, Babeș-Bolyai University, Cluj, vol 1. 23–24 Mar 2012, pp 445–452
Mihu-Pintilie A, Romanescu G, Stoleriu CC, Nicu IC, Asăndulesei A, Schmaltz E (2014) Natural dam lakes from Cuejdiu watershed (Stânișoarei Mountains)—non–invasive methods used for bathymetric maps. In: Gâştescu P, Marszelewski W, Breţcan P (eds) Proceedings conference of the water resources and wetlands, vol 1. Tulcea, 11–13 Sept 2014, pp 130–137
Mihu-Pintilie A, Asăndulesei A, Nicu IC, Stoleriu CC, Romanescu G (2016) Using GPR for assessing the volume of sediments from the largest natural dam lake of the Eastern Carpathians: Cuejdel Lake, Romania. Environ Earth Sci 75:710. https://doi.org/10.1007/s12665-016-5537-1
Mîndrescu M, Iosep I, Cristea IA, Forgaci D, Popescu DA (2010), Lacurile Iezer şi Bolătău (Obcina Feredeului)–cele mai vechi lacuri de baraj natural formate prin alunecare din România. In: Gâştescu P, Breţcan P (eds) Lucrările Simpozionului Naţional–Resursele de apă din România, Târgovişte, vol 1. 11–13 June 2010, pp 272–281

Mîndrescu M, Cristea A, Hutchinson SM, Florescu G, Feurdean A (2013) Interdisciplinary investigations of the first reported laminated lacustrine sediments in Romania. Quat Int 293:219–230
Minea I, Romanescu G (2007) Hidrologia mediilor continentale–Aplicaţii practice. Demiurg Publishing House, Iaşi
Montandon F (1933) Cronologie des Grands Eboulements Alpins du debut de l'ère chrètienne à nos jours. Mat pour l'étude des calamités. Société de Géographie (Genève) 32:271–340
Morariu T (1946) Câteva observaţii asupra crovurilor din Banat. Rev Geogr 2(4):37–50
Morariu T. Savu A (1960) Les types génétiques de lacs et leur répartition sur le territoire de la RP Roumaine, vol 4. Revue de Géologie et de Géographie. Bucharest
Mrazec L (1901) Lacs salés de la Roumaine. In: Archiv Sci Physiq Nat (Archives of the physical and natural sciences), Généve
Murgoci G (1907) La plaine Roumaine et le Balta du Danube. Paper presented at the 3rd international symposium on the accounts of International Congress of Oil, Bucharest
Murray J, Pullar F (1910) Bathymetrical survey of the Scottish fresh-water lochs. Challenger Office, Edinburgh
Năstase IG (1949) Un lac necunoscut: Lacul Bălătău. Rev Sci Vasile Adamachi 21(1–2)
Năstase A (1960) Lacurile Capra şi Căpriţa din Masivul Făgăraşului. Probleme De Geografie 7:267–274
Noord H, Phuntsho T (2005) Hazard Assessment of Two Slides Areas in the Kharthi Ri Chu catchment Ranjung. Trashigang Dzongkhag, East–Bhutan. DGM Report, March 2005
Orghidan N (1932) Urme de gheţari pe Siriu. In: Bulletin of Romanian Royal Society of Geograph, vol 51. Bucharest
Pandi G, Buzila L (2004) Caracteristici hidro–geomorfologice ale sedimentarii in Lacul Rosu. In: Geography within the Context of Contemporary Development, Cluj-Napoca University Press, Cluj
Pandi G, Magyari ZS (2004) Realizarea hărţilor batimetrice pe calculator. Modelul Lacul Roșu. Stud Univ Babeș-Bolyai Cluj 1:55–60
Petrescu P (1932) Etude geochimique des eaux des limans du SE de la Bessarabie. An Inst Geol Rom 17:225–263
Pettinga JR (1987) Ponui Landslide: A deep-seated wedge failure in Tertiary weak-rock flysch, Southern Hawke's Bay, New Zealand. J Geol Geophys 30(4):415-430
Pişotă IT (1956) Câteva observaţii hidrologice asupra lacului Bâlea şi bazinului hidrografic Bâlea-Cârţişoara. Revista Natura 8:40–47
Pişotă IT (1957) Câteva observaţii hidrologice asupra lacului Urlea. Anal Univ CI Parhon-Sci Nat 27:253–264
Pişotă IT (1958) Lacurile de pe flancul nordic al Masivului Făgăraş. Stud Cerc Geo Buc 1:70–87
Pişotă IT (1971) Lacurile glaciare din Carpaţii Meridionali. Romanian Academy Publishing House, Bucharest
Plafker G, Ericksen GE, Fernandez Concha J (1971) Geological aspects of the May 31, 1970, earthquake. Seism Soc America Bull 61(3):543–578
Plăvan G (2010) Cercetări privind structura şi dinamica comunităţilor macro–zoobentosului în lacul de acumulare Izvorul Muntelui-Bicaz. Alexandru Ioan Cuza University of Iaşi, Disertation
Pourriot R, Meybek M (1995) Limnologie générale. Masson, Paris
Pusharenko VP (1982) Regularities of mud-rock flow formation in central Asia. In: Sheko A (ed) Landslides and mudflows. Centre of International Projects, GKNT, Moscow, USSR, Alma-Ata International Seminar, October 1981, Reports 87-106
Rădulescu I, Năstase A (1960) Observaţiuni limnologice asupra limanului maritim Tatlageac. Anal Univ CI Parhon–Sci Nat 23
Rădoane N (2002) Un nou lac de baraj natural în bazinul Bistriţei Moldoveneşti-Lacul Cuejdel. Stud Cerc Geograph 49:76–82
Romanescu G (2000) Limnologie şi Glaciologie-Partea I Limnologie. Suceava University Publishing, Suceava

Romanescu G (2008) Inventarierea şi tipologia zonelor umede şi apelor adânci din Podişul Moldovei (The inventory and the typology of wetland areas and the deep waters of Moldavian Plateau). Terra Nostra, Iaşi

Romanescu G (2009) Trophicity of lacustrine waters (lacustrine wetlands) on the territory of Romania. Lakes Reservoirs Ponds 3:62–72

Romanescu G (2010) Caracteristicile fizico-chimice şi tipologia zonelor umede din Grupa Centrală a Carpaţilor Orientali. In: Gâştescu P, Breţcan P (eds) Lucrările Simpozionului Naţional–Resursele de apă din România, vol 1. Târgovişte, 11–13 June 2010, pp 33–39

Romanescu G, Stoleriu C (2010) Parametrii morfobatimetrici ai cuvetei lacustre Roşu (Hăghimaş). In: Gâștescu P, Brețcan P (eds) Resursele de apă din România–Vulnerabilitatea la presiunile antropice, vol 1. Târgovişte 11–13 June 2010, pp 308–314

Romanescu G, Stoleriu C, Enea A (2013) Limnology of the Red Lake, Romania. An interdisciplinary study. Springer, Dordrecht

Rudoy AN (2002a) Glacier–dammed lakes and geological work of glacial superfloods in the Late Pleistocene, Southern Siberia, Altai Mountains. Quat Int 87:119–140

Rudoy AN (2002b) Late-Quaternary diluvial floodstreams in the mountains of Altai and Tuva-paleoglaciological base, geological effect, level of problem study. Lunar Planet Sci 33:1109–1110

Rusu C, Mărgărint MC, Rusu E, Boamfă I (2002) Noi arii de interes ecologic din Moldova: Parcul Forestier Vânători – Neamţ şi Lacul Crucii (bazinul hidrografic al Cuejdelului), vol 1–2. Revista Terra, Bucharest

Ruttner F (1963) Fundamentals of limnology, 3rd edn. English edition (trans: Frey DG, Frey FEJ). Universuty of Toronto Press, Toronto

Săndulache A (1970) Lacurile dulci din Câmpia Transilvaniei. Disertation, Pedagogical Institute of Oradea

Schram M (1970) Studiul hidrologic al lacurilor din Câmpia Moldovei. Disertation, Babeș–Bolyai University of Cluj–Napoca

Schuster RL (1985) Landslide dams in the western United States. IVth International Conference and Field Workshop on Landslides-Tokyo, Proceedings 1:411–418

Schuster RL, Costa JE (1986) A perspective on landslide dams. In Schuster RL (ed), Landslide dams-Processes, risk, and mitigation, American Soc Civil Engin Geotech Special Pub 3:1–20

Schneider JF, Gruber FE, Mergili M (2011) Recent cases and geomorphic evidence of landslide-dammed lakes and related hazards in the mountains of Central Asia, In: Proceedings of the second world landslide forum, vol 1. Rome, 3–9 October 2011, pp 1–7

Schuster R., Evans SG (2011) Engineering Measures for the Hazard Reduction of Landslide Dams. In: Evans SG, Hermanns RL, Strom A, Scarascia-Mugnozza G, (eds) Natural and artificial rock-slide dams. Lecture Notes in Earth Sciences, vol 133. Springer, Heidelberg, p 77–100

Schuster RL, Wieczorek GF (2002) Landslide triggers and types. In: Rybar J, Stemberk J, Wagner P (eds) Landslides–proceedings of the 1st European conference on landslides, vol 1. Prague, 24–26 June 2002, pp 59–79

Scott KM (1985) Lahars and lahar–runoff flows in the Toutle–Cowlitz River system, Mount St. Helens, Washington—Origins, behavior, and sedimentology. US Geological Survey Open-File Report 85-500

Silva L, Cocheme JJ, Canal R, Ouffield WA, Tilling RI (1982) The March-April 1982 eruptions of Chichonal Volcano, Chiapas, Mexico-Preliminary observations. EOS 63:11–26

Smith IR (1975) Turbulence in Lakes and Rivers. SP 29, Freshw Biol Assoc, p 79

Şenchea N (1948) Formarea lacurilor din România. Rev Sci Vasile Adamachi 34(2)

Ştefănescu G (1888) Notă asupra secărilor Lacului Sărat. In: Anuarul Biroului Geologic (Annual Proceeding of Romanian Geological Office), vol 5. Bucharest

Swanson FJ, Graham RL, Grant GE (1985) Some effects of slope movements on river channels. In: Proceedings of the international symposium on erosion, debris flow, and disaster prevention, vol 1. Tsukuba, Japan 3–5 Sept 1985, pp 273–278

Swanson FJ, Oyagi N, Tominaga M, (1986) Landslide dams in Japan. In: Schuster RL (ed) Landslide dams—processes, risk, and mitigation, vol 3. American Society of Civil Engineers Geotechnical, Special Publication, pp 131–145

Tiner RW (1999) Wetland Indicators–a guide to wetland identification, delineation, classification, and mapping. Lewis Publishers, Boca Raton, London, New York, Washington DC

Thorarinsson S (1953) Some new aspects of the Grimsvotn problem. J Glaciol 2:267–275

Ujvari I (1959) Hidrografia R.P. România–subcapitolul Lacuri. Romanian Academy Publishing House, Bucharest

Ujvari I, Găştescu P (1958) Evaporaţia apei de la suprafaţa lacurilor din R.P. România. Mete Hidro Gospodărirea Ape, Bucharest

Varnes DJ (1978) Slope movement types and processes. In Schuster RL, Krizek RJ (eds), Landslides-Analysis and control: Transportation Research Board, National Academy of Sciences, Washington, Special Report 176:11–33

Vuichard D, Zimmermann M (1987) The 1985 catastrophic drainage of a moraine-dammed lake, Khumbu Himal-Cause and consequences. Mt Res Dev 7:91–110

Weischet W (1963) Further observations of geologic and geomorphic changes resulting from the catastrophic earthquake of May 1960 in Chile. Bull Seism Soc Am 53(6):1237–1257

Woolley RR (1946) Cloudburst floods in Utah, 1850-1938. U.S. Geol Survey Water Supply Paper 994:1–128

Wu J (1986) Landslide dams in south-central China. In: Schuster RS (ed) Landslide dams—processes, risk, and mitigation, vol 3. American Society of Civil Engineers Geotechnical, Special Publication, pp 146–162

Yesenov UY, Degovets AS (1979) Catastrophic mudflow on the Bol'shaya Almatinka River in 1977. Sov Hydrol 18:158–160

Chapter 3
Geographic Location and Boundaries of Study Area

Abstract The natural dam lake Cuejdel is situated in the Central Group of the Eastern Carpathians (NE Romania), within Stânișoarei Mountains, part of the Moldavian–Transylvanian Carpathians group. They correspond to the eastern sector of the External Marginal Syncline, with a median position in the flysch chain. The southern subunit of the Stânișoarei Mountains—Neamț Mountains is positioned between Bistrița Valley in the west and Cracău-Bistrița Depression in the east (Moldavian Subcarpathians). The lake was formed in 1991, in the upper hydrographic basin of Cuejdiu, one of the left tributaries of the Bistrița River. The name Cuejdel Lake was given after the hydronym of the brook on which it formed. Etymologically, this word comes from Slavonic (Chivejdi), which means "dry valley" or "river stone". It is also known as the Cross Lake (in Romanian—Crucii Lake). The most important tributaries are the Pârâul Rotund, Cuejdel, Glodu and Făgetu brooks. The major orographic milestones from the lake basin catchment are Muncelului Peak (1077 m), Tarnițelor Peak—Gârcina Massif (1081 m) and Măties Summit (986 m).

3.1 The Mathematical Position and Physical-Geographical Boundaries

The natural dam lake Cuejdel, locally known as Cross Lake (in Romanian–Crucii Lake), is situated south-east of the Stânișoarei Mountains, a component part of the Moldavian-Transylvanian Carpathians, a subdivision of the Eastern Carpathians (North-Eastern Romania). The lacustrine basin was formed in the superior hydrographic basin of Cuejdiu, a tributary of the Bistrița River, at about 1.5 km upstream in the Cuejdel brook, the left tributary of the Cuejdiu River. It originally formed due to a large landslide that caused rock displacement on the left slope of Muncelu Peak (1077 m) during the summer of 1991. The latitude and longitude coordinates are situated between 47°01′54″ N in the southern sector and 47°02′21″ N in the northern one, and between 26°13′02″ E in the western sector and 26°13′07″ E in the south-eastern one (Fig. 3.1).

Territorially, the study area belongs to the territory of Gârcina commune in Neamț County, being situated at just 3 km away from the southern boundary of Crăcăoani

A. Mihu-Pintilie, *Natural Dam Lake Cuejdel in the Stânişoarei Mountains, Eastern Carpathians*, https://doi.org/10.1007/978-3-319-77213-4_3

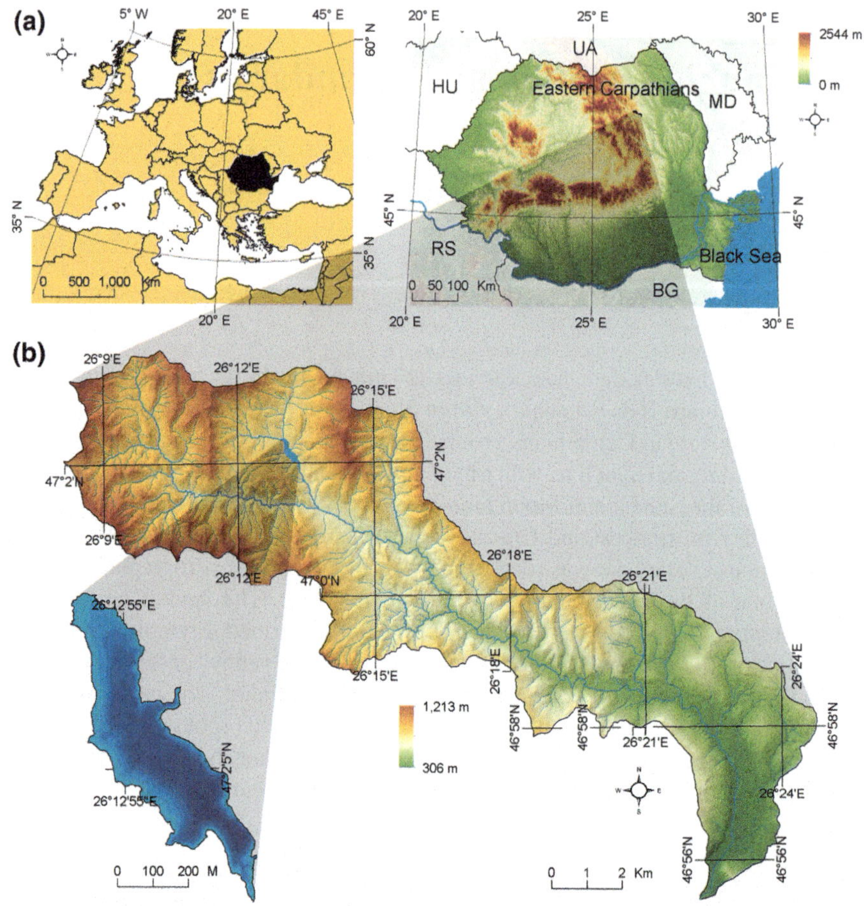

Fig. 3.1 **a** Geographic location of study area in the Eastern Carpathians (Romania); **b** The mathematical position of Cuejdel lake in the upper hydrographic basin of the Cuejdiu river

commune. The nearest localities are: northwards, Cracăul Negru (4.5 km) and Magazia (8 km), and south-eastwards, Cuejdiu (8.5 km), Gârcina (12.5 km) and Piatra Neamț town (21 km).

Due to the morphological particularities in which the flooded valley evolved, the current lacustrine basin inherited a long shape with a north-west—south-east direction. The water level was constant and situated at an absolute altitude of 665.5 m until 1994–1995, when more flash floods occurred and deepened the drainage lake channel. The relative altitudine at which it is found today is approximately 661 m (Ichim et al. 1996; Rădoane 2002; Mihu-Pintilie and Romanescu 2011; Mihu-Pintilie et al. 2012).

The main hydrographic milestones of the lake are: Cuejdel brook confluency with Glodu brook in the upper sector of the lake (47°02′18″ N, 26°12′05″ E), the outlet

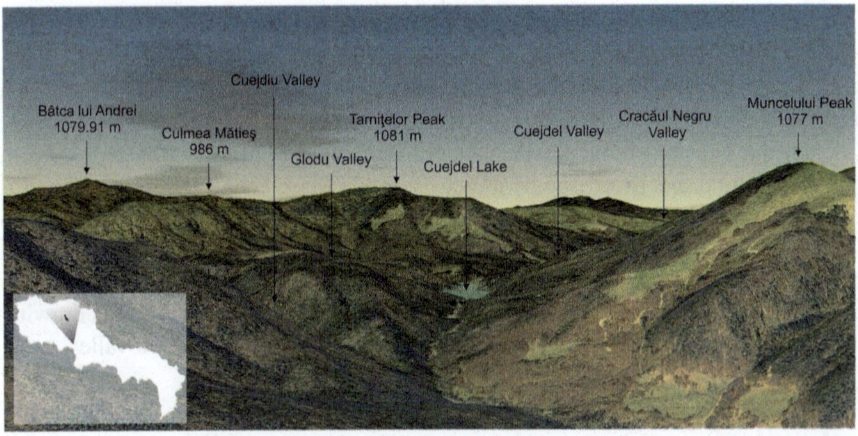

Fig. 3.2 The orographic major milestones from the lake basin catchment—northwest view in the upper basin of Cuejdel river

of Făgetu brook from the left versant (47°02′13″ N, 26°13′03″ E), and of Crucii brook, from the right versant (47°02′37″ N, 26°12′51″ E), both positioned in the middle part of the lacustrine unit and of the outlet of Rotund brook, located in the lower sector, on the right versant. The orographic landmarks that dominate the area in terms of altitude are mostly a component part of the watershed delimiting the catchment basin of the lake (8.75 km^2). This limit is composed of the alignment: Piciorul Rotund at south-west (822 m), Coasta Mătieș (986 m) at west, Tarnițelor Peak (1081 m), Gârcina Massif (873 m) and Dealul Măgurii (1117 m) at north, and the summit and Muncelul Peak (1077 m) at east (Fig. 3.2).

As a part of the hydrographic system of Cuejdiu River, the integration of the Cuejdel brook is considered necessary at the level of this hydrographic basin. The geographic coordinates that define the area are: 47°03′47″ N—northern boundary; 46°55′22″ N—southern boundary; 26°08′26″ E—western boundary and 26°24′54″ E—eastern boundary. Its surface is of 97.41 km^2 and develops a perimeter of 64.75 km. The length of the main course is about 29.11 km with an altitudinal interval between 305 m in the Bistrița Corridor and 1213 m in the Prislopașu Peak.

The watershed starts from south to north, respectively from Bistrița Valley towards Dealul Gospodina (419.9 m) and Dealul Cozla (676 m), from where it presents a south-east—north-west direction, passing through several important summits: Prohodiștea Peak (805 m), Bâtca lui Ivan (854.5 m), Dealul Mare (1053.2 m), Arinilor Peak (957 m), Gherman Peak (1053 m), Bâtca Vaca Babei (1104.5 m) through Prislopașu Peak (1213 m). Henceforth, the limit is heading north again, passing through Burlacilor Peak (1056 m), ending under Buhalnița Peak at a relative altitude of 1027 m. This interfluvial alignment separates southwards Cuejdiu Valley from Bistrița Valley and south-eastwards Cuejdiu Valley from the basins of Valea Mare, Pângărăcior and Buhalnița streams.

The transition to the left slope of Cuejdiu basin continues from west to east through: Bâtca lui Andrei (1079.9 m), Tarniţelor Peak (1081 m), Gârcina Massif (873 m), Dealul Măgurii (1117 m), towards Dealul Leaturi (812.2 m), forming the watershed between the upper Cuejdiu river basin, composed by Cerbul, Argintăria, Cuejdel and Ţiganca brooks and the valleys of Bouleţ and Cracăul Negru rivers. From this point, the peaks divert southwards to Ţiganca Mare Peak (957 m), then south-eastwards, crossing several interfluves with decreasingly lower altitudes heading to Dealul Balaurului (522.9 m). This sector represents the limit towards Almaş stream basin to the northward and Ciocoiţa brook to the north-eastward. The watershed ends in Pietricica Peak (527.8 m) and again in Bistriţa Valley, after heading south, drawing the boundary between the Cuejdiu lower sector and the Frăsinel brook valley.

3.2 Regional Integration Elements and Territorial Units

Situated at the contact between the Carpathian area and Moldavian Subcarpathians, Cuejdiu Valley drains sizeable surfaces within two different landform subunits: the Neamţ Mountains (74.1 km^2), throughout the middle and upper course, as well as on the western side of the lower sector and the Cracău-Bistriţa Depression (22.91 km^2), on the lower course, but with a higher extension on the north-eastern and eastern side. After crossing the narrow sector located between Dealul Gospodina (419.9 m) and the Pietricica Ridge (527.8 m), very close to Cuejdiu outlet, the valley corresponds to Bistriţa Corridor on a sector totalling only 0.47 km^2.

The delimitation of physical-geographical subunits from the Cuejdel Lake area and the Cuejdiu hydrographic basin was realized based on the geological (lithostratigraphical structure, sediments nature), geomorphological (hypsometry, depth and relief fragmentation, declivity, slope orientation), and on the biopedogeographic criteria (forested areas distribution, soil layer characteristics). In regards to those three areas, these criteria make a major difference for both the relief conditions which have evolved, and the geomorphological features that represent the triggering premise of extensive gravitational processes and lacustrine accumulations through landslide dam processes.

The subunit of the Neamţ Mountains corresponding to the Cuejdiu Valley grafted on the south-eastern slope of the Stânişoarei Mountains, from where the main physical-geographical characteristics have arisen: complex lithologic structure corresponding to the external flysch and to the contact area with the Pericarpathian unit: Anticlinal fold Doamna—Horaiţa, False anticlinal Almaşu; well-defined slopes and ridges on the tectonic interference lines: Tarniţelor Peak (1081 m)—Coasta Mătieş (986 m), Dealul Măgurii (1117 m)—Muncelului Peak (1077 m); altitudes between 400 and 1200 m: maximum altitude in Prislopaşu Peak (1213 m); relief energies between 200 and 400 m; a hydrographic network formed especially through selective modelling: Runcu, Cerbului and Ţiganca valleys; a large spread of gravitational processes like landslides, subsidences and rock rolling. This geomorphological context concerns not only the formation of the natural dam lake Cuejdel, but also the two other smaller lakes, both with the same genesis of lacustrine basins.

Climatically, this sector corresponds to the larger area of the temperate continental regime with excessive influences due to south-eastern exposure of the valley towards the Subcarpathian area. Nevertheless, compared to the other physical-geographical subunits within the Cuejdiu basin, there are some differences both in the thermal gradients that vertically decrease due to higher altitudes, and in the annual rainfall amounts, that are considerably higher, due to the mountain peaks acting as an orographic barrier.

From the perspective of the biogeographical conditions, the major characteristic is defined by the very high degree of coverage with forest vegetation (>75%), where the main wood species are spruce, beech and oak, grouped both in compact units, and in batches of mixed forests. As a consequence of the high degree of afforestation, the soils are mainly formed of Cambisols (Eutric cambisol, Dystric cambisol), intercalated with Luvisols and sometimes with undeveloped soils formed from the superficial deposits.

The anthropic component is absent in the middle and upper sectors of the hydrographic basin, and in the lower third, the Cuejdiu village focuses on the first terraces of the river, advancing up to the confluence of the Cuejdiu and Țiganca brooks. Only at the contact with the Subcarpathian area does the valley widen considerably, an aspect which also determines an expansion of human settlements on the upper terraces of the river.

The subunit of the Cracău-Bistrița Depression detaches in the great Pericarpathian unit as a depressionary sector conducted by the Gârcenița Valley leading to contact with the mountainous area. The entire region is considered to be a consequence of permissive lithology, mostly clay, shaped in the past by a series of rivers flowing in a north-west—south-east direction. After the Cuejdiu stream capture by the Bistrița River, west from Pietricica Peak, the depressionary corridor was redirected southwards, practically returning to the mountainous area (Lupașcu 1996).

The main features are represented by the reduction of forested areas and the diversification of edaphic coverage, especially through the emergence of Cernisols (phaeozems). This has led to a considerable increase in land used for agricultural purposes and of areas occupied by pastures and meadows. Also, low altitudes and low relief energy facilitated the almost circular disposition of Gârcina village at the confluence of the Cuejdiu water course with the Gârcenița brook. Downstream, due to the maximum extension of Cuejdiu Valley, the main forms of urban organization appear and are represented by Dărmănești district, a component part of Piatra-Neamț town.

The subunit of the Bistrița Corridor occupies the smallest surface within the studied area (0.47 km^2), but detaches from the other subunits by the specific morphology of Bistrița Valley. The plane relief from Pietricica and Cozla peaks base is entirely occupied by the urban area where the geographical landscape suffered important changes. In terms of hydrotechnical planning, the lower watercourse was embanked, which represents an important measure to protect the city against floods.

References

Ichim I, Rădoane N, Rădoane M (1996) Procese geomorfologice cu intervale de recurenţă mare în arealul munţilor flişului. Exemplificări din Judeţul Neamţ (Geomorphological process with high frequences in flysch mountains area. Examples from Neamț County). In: Studii şi cercetări—extras 8, Muzeul de Ştiinţe Naturale Publishing, Piatra-Neamţ, p 15

Lupașcu G (1996) Cracău-bistrița depression. A pedogeographic study. Corson Publishing House, Iași (in Romanian)

Mihu–Pintilie A, Romanescu G (2011) Morphometric and morphological suitability of the relief from the crucii (Cuejdel) lake basin (Stânişoarei Mountains). In: Pandi G, Moldovan F (eds) Proceedings of the air and water components of the environment, vol 1 Babeș–Bolyai University, Cluj, 18–19 March 2011, 305–313

Mihu-Pintilie A, Romanescu G, Stoleriu C (2012) Morpho–bathymetric parameters of recess crucii (Cuejdel) lake (Stânişoarei Mountains). In: Pandi G, Moldovan F (eds) Proceedings of the air and water components of the environment, vol 1, Babeș–Bolyai University, Cluj, 23–24 March 2012, 445–452

Rădoane N (2002) Un nou lac de baraj natural în bazinul Bistriţei Moldoveneşti-Lacul Cuejdel. Stud Cerc Geogr 49:76–82

Chapter 4
Physical-Geographical Conditions of the Cuejdiu Valley and Their Importance in the Formation of Cuejdel Lake

Abstract It is necessary to integrate the Cuejdel Brook and the catchment areas of Cuejdel Lake as a part of the hydrographic basin of Cuejdiu River. The analysis of physical-geographical conditions from the Cuejdel Valley and their role in the formation of Cujdel Lake emphasize a number of factors that led to the destabilization of the versants and generated the slide complex. It is important to mention the role of: geological factors (flysch layers with clay and argillaceous-marl deposits and compact sandstones), relief (relief energy >300 m, declivity 30°–50°), rainfall regime (cumulated pp. 05–07. 1991—741.4 mm), hydrography (Order IV—Horton Strahler classification) and the structure of valleys (transversal/longitudinal on the general direction of lithologic layers), bioedaphic cover (deciduous, coniferous and mixed forests—96.2%), and last but not least, the anthropic factor (the density of forest exploitation road network 10.67 km/km^2). The list of factors that led to landslides can be completed by the earthquake from 1990 (5.4° Richter).

4.1 The Role of Geological Factors

The geological database was produced as a result of consulting the literature, from which mention should be made of the research reports realized between 1958 and 1968 and conducted by specialists of the Romanian Geological Institute. Completing reports of the unexplored places was made through field observations and by using geological sketches (1:50,000) performed for the Cuejdiu Valley by Mirăuță and Mirăuță (1964). As a general rule cartographic support was provided by the use of the geological map 13–Piatra Neamț (ed. 1968), scale 1:200,000, produced by the Romanian Geological Institute based on geological maps: 1:50,000 (ed. 1958); 1:10,000 (ed. 1959); 1:50,000 (ed. 1962); 1:5000 (ed 1964).

Thus, the upper hydrographic basin of the Cuejdiu river, the natural collector of the catchment basin of Cuejdel Lake, mostly overlaps with an area characterized by a very complex lithology and a pronounced tectonics that corresponds to the parautohton marginal flysch of the external unit from Eastern Carpathians. More specifically, it developed in the middle sector of Bistrița-Râșca subunit within Vrancea Nappe, where it discharges the flysch formations over the Subcarpathian area in a west-east

A. Mihu-Pintilie, *Natural Dam Lake Cuejdel in the Stânişoarei Mountains, Eastern Carpathians*, https://doi.org/10.1007/978-3-319-77213-4_4

direction. Due to this geological conjuncture, two major lithologic areas can be identified within the hydrographic basin: a middle-superior one, structurally dominated by the petrographic complex of the external flysch, the context in which the Cuejdel Lake formed, and an inferior one, much more limited, predominantly consisted of molasse deposits (Fig. 4.1).

4.1.1 Lithological Structure and Stratigraphy

The lithostratigraphical structure of the flysch area begins in Cuejdiu hydrographic basin with the formations of the early and late Cretaceous periods that are a characteristic of Vrancea Nappe and only with the late Cretaceous period within Tarcău Nappe. The sequence continues with Palaeocene–Eocene formations, also characteristic of Vrancea and Tarcău units, the last of which being less represented. Over these formations takes place the emergence of others of the Oligocene and Miocene ages, this time, characteristic only for Vrancea Nappe. The last lithologic term from the hydrographic basin of Cuejdiu appears in the lower basin and is a characteristic of the middle and late Pleistocene periods, through diluvial-proluvial deposits or are represented by a series of deposits of the Holocene age, insularly arranged along riverbeds.

Cretaceous formations from Vrancea Nappe. The Audia formation from Vrancea Nappe (ne-al), were originally known by the names Stratele de Sărata (Băncilă 1955) and Stratele de Streiu (Dumitrescu 1952) respectively, but the current name comes about due to the stratigraphic interval similar to the one from the Audia and Tarcău unit (Joja et al. 1968; Mutihac and Ionesei 1974; Grasu et al. 1988). Within the Bistriţa–Râşca subunit of Vrancea Nappe, the formation is related to the Doamna–Horaiţa anticlinorium structure, where the stratigraphic sequence is composed of three horizons: a clay and limestone basal member, an average member of clay minerals and a superior member of limestones with silicification (Mirăuţă 1962; Mirăuţă and Mirăuţă 1964).

In the middle sector of the Cuejdiu hydrographic basin corresponding to the western flank of Doamna–Horaiţa anticlinorium, all three of the terms mentioned terms, in contract to the eastern flank, where the diapir character frequently interrupts this sequence. Thus, not only on the right slope of Ţiganca creek, but also from its confluency with the Cuejdiu river and up to beyond the settlement of the same name, the alternation of the three terms is preserved, defining the area as one of stratotype for Audia Formation from Vrancea Nappe (Dumitrescu 1952; Filipescu et al. 1962; Mirăuţă and Mirăuţă 1964; Grasu et al. 1988).

The formation of striped clays (vr-tu) occurs after the quartz arenite episode or limestones with silicification that ends the sedimentation in Audia—Sărata facies, when the accumulation of red, green and black clays takes place. This aggregate is known under different names, also representing a very good reference point of lithologic correlation from one nappe to another and being frequently formed from sandstones, marlstones and limestones, with clayey intercalations. In the Vrancea

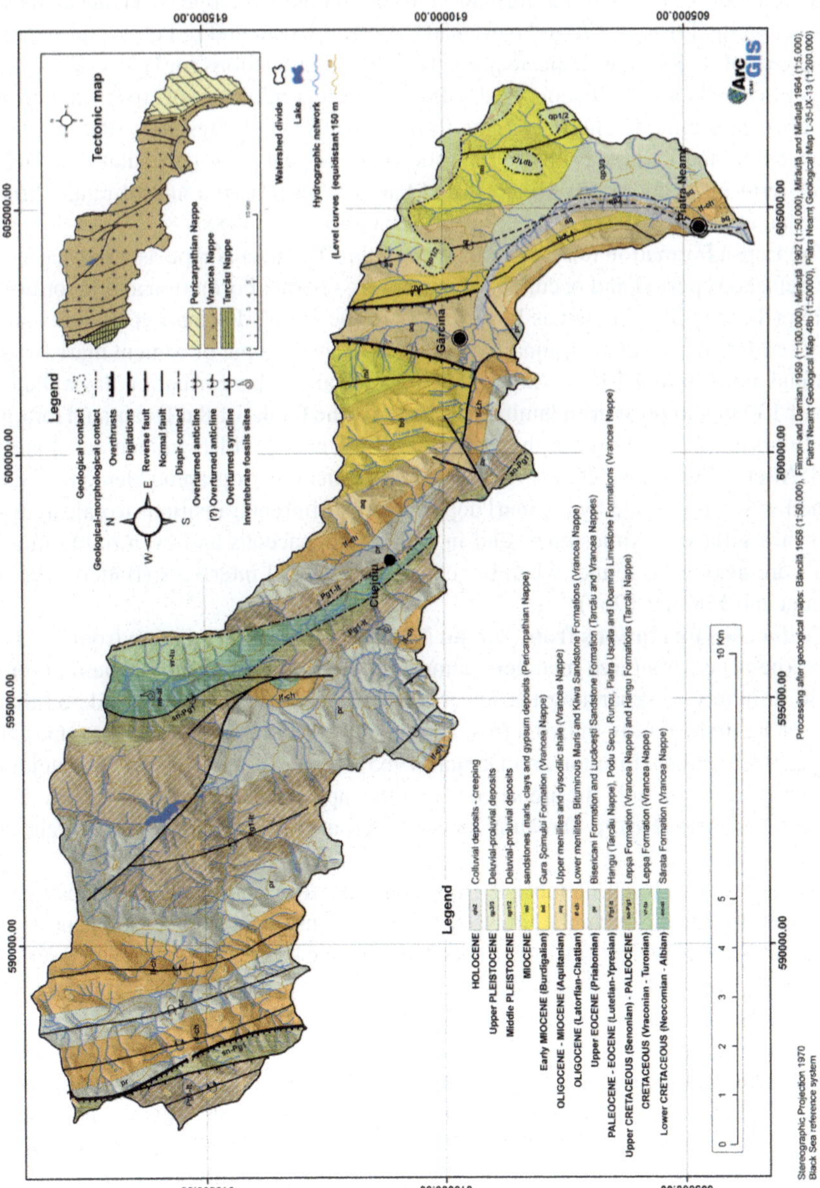

Fig. 4.1 Geologic and tectonic map of Cuejdiu hydrographic basin

unit the formations are known as Stratele de Cârnu–Șiclău (Băncilă and Hristescu 1963), striped clays (Mirăuţă and Mirăuţă 1964) or as Stratele de Tisaru (Dumitrescu 1952).

Within the studied area, the formation occurs on the left slope of Ţiganca creek, disposed in the form of a 20 m layer, made of black, green and red clays, in successive layers of 2–6 m, that indicate green radiolite intercalations, in layers of 2–5 m, and occasionally black siliceous rocks. From a petrographic perspective, a series of ferrous rocks similar to cherts, ranging from jasper to radiolarites, as well as argillaceous, silica and carbonaceous pyroclastic rocks appear with particular reference to this term in a clay matrix (Băncilă and Papiu 1962; Mirăuţă and Mirăuţă 1964; Idriceanu et al. 1967; Papiu et al. 1967, 1983; Grasu et al. 1988).

The Lepșa Formation (sn-Pg1) is known within Vrancea Nappe as a last term of the Cretaceous period and occurs between Audia—Sărata Formation and Izvor Formation (Runcu), also known as lower layers of the Hangu Formation (Băncilă and Hristescu 1963), or can occasionally be found under the name of cement marlstones (Mirăuţă and Mirăuţă 1964). Along the Cuejdiu Valley, this formation has a thickness of 130 m and presents a laminated aspect on the flanks of the Doamna–Horaiţa anticlinorium, particularly on the upper side of Ţiganca Hill (Mutihac and Ionesi 1974; Micu 1979; Grasu et al. 1988). The predominant petrographic element of the formation is represented by the marl deposits with different transitions to calcareous clays or argillaceous limestones, and in addition arenaceous and even rudite limestones are also to be found, when the epiclastic material interferes (Băncilă 1958; Mirăuţă and Mirăuţă 1964).

Cretaceous formations from Tarcău Nappe. The Audia Formation from Tarcău nappe (ne-al) covers three members, almost similar to those in the Audia unit, but due to difficulty in separating the lower and middle member, it is considered that they can be divided only in two terms: an argillaceous basal member and a superior arenaceous member (Filimon and Damian 1965). In the upper sector of the Cuejdiu hydrographic basin, only the arenaceous member appears as a last term of the Audia Formation sequence from Tarcău Nappe, located at the interference with Vrancea Nappe.

The arenaceous basal member is formed mostly from massive sandstones (several meters), presenting parallel muscovite laminations on the stratification surfaces and an intense geological fracture. Subordinately, black clays emerged with a series of blackish-grey rocks with a sublithographic limestones aspect, considered argillaceous marlstones (Filimon and Damian 1965; Grasu et al. 1988).

Hangu Formation (sn-Pg1) represents a distinct lithostratigraphical entity that describes the Senonian facies from the central and eastern part of Tarcău Nappe and partially of Vrancea Nappe, initially known under the name of Strate cu inocerami (Macovei and Atanasiu 1925; Macovei et al. 1927) or Strate de Hangu (Atanasiu 1939) in the literature. In the area of the Cuejdiu hydrographic basin, these layers appear on a relatively restricted strip as a continuation of the superior arenaceous member of the Audia Formation from Tarcău Nappe, their dating being based on the presence of inoceramus and ammonite fossils or microfaunistic and palynological associations (Joja and Chiriac 1964; Ionesi and Tocorjescu 1968; Turculeţ 1971; Ştefan 1974; Olaru 1978).

Palaeocene–Eocene formations from Vrancea Nappe. The Izvor Formation (Runcu) (Pg1-lt) is the first entity that marks the beginning of the Palaeocene within the external flysch area and is situated between Hangu Formation and Lepşa Formation at the base, respectively Straja and Piatra Uscată Formations at the top. For this reason, another term used to define them is Stratele de Hangu superioare (Joja and Chiriac 1964), or Stratele de Runcu–Putna (Micu 1979). Within the studied area, this formation emerges completely open on Runcu creek, a right tributary of Cuejdiu, where the following petrographic types can be distinguished: biosparites and sparites with bioclasts and green rocks extraclasts, siltitic micrites, siltitic marls and clays (Ionesi 1971; Grasu et al. 1988).

Piatra Uscată Formation (Pg1-lt) is a continuation of Straja striped flysch from Tarcău Nappe that preserves its particularities for Vrancea Nappe, located between Izvor Formation and Jgheabul Mare Formation. In the analysed area, it emerges on the riverbank of Runcu creek and in some sectors of Cuejdiu, longitudinally arranged on the western flank of Doamna–Horaiţa fold. The formation includes as petrographic terms biosparites or arenaceous biosparites, quartz arenites, gaize-spongolite, clays and silica clays (Dumitrescu 1952; Băncilă and Hristescu 1963; Mirăuţă and Mirăuţă 1964; Mutihac and Ionesi 1974; Micu 1979).

Jgheabul Mare Formation (Pg1-lt) develops exclusively within the Bistriţa–Râşca subunit connected to the western flank of Doamna–Horaiţa anticlinorium, with the Piatra Uscată Formation at the base and supporting Doamna Limestones (Olteanu 1952, 1953; Mirăuţă 1969; Micu 1979). Despite the fact that the stratotype section is situated outside the analysed area, this term can be identified in a stratigraphic manner within some sectors of the Cuejdiu Valley. The rocks from this formation present a high hardness, well stratified in centimetric or decimetric banks, with a massive aspect due to the total absence of loam intercalations. The petrographic hybrid terms indicate a carbonate sedimentation associated with organogenic silica intake and episodically, the epiclastic, quartz material (Grasu et al. 1988).

The Doamna Limestone Formation (Pg1-lt) is situated between the Jgheabul Mare Formation at the base and the Bisericani Formation in the upper part, and it is specific to the Vrancea unit and the external limit of Tarcău Nappe, which places it in the middle Eocene (Joja and Manoliu 1970; Ionesi 1971). Within the Bistriţa–Râşca subunit, specifically in the cross section of the Doamna creek, this formation presents the following types of rocks under a petrographic report: silicolitic biomicrites, arenaceous biosparites, calcareous sandstones, silt and marlstones (Joja 1952; Papiu 1960; Ionesi 1971).

Bisericani Formation (pr) represents the equivalent of the Podu Secu and Plopu Formations from Vrancea Nappe, the Bistriţa–Râşca subunit, developed in the external lithofacies of Tarcău and Vrancea Nappes and is stratigraphically arranged between the Doamna Limestones at the base and Lucăceşti sandstone at the top (Athanasiu et al. 1927; Botez et al. 1967). The formation includes a basal member of red and green clays mixed with sandstones, a medium member of greyish-green clays and a superior member of globigerina marlstones. The stratotype area described in the literature appears on Ştirbosu creek that divides the eastern flank of the Doamna–Horaiţa anticlinorium, indicating lithostratigraphical similarities with the western flank. Thus, in the area of the Cuejdel hydrographic basin, the formation corresponds to Palaeocene–Eocene deposits that emerge throughout the middle and middle-superior sectors of the study area.

Clay rocks at the base belong to proper clays and slightly greyish-green clays categories, fact that gives them a hybrid status among clays and silts. Occasionally, in the middle member of greyish-green clays can be found rare and discontinuous intercalations of not only sandstones and breccias with green clays elements, but also siderite limestone lentils. Globigerina marlstones from the upper part of the column do not present a constant presence, but a deficiency in the lithostratigraphical alternation of this formation (Săndulescu and Săndulescu 1964; Mirăuţă and Mirăuţă 1964).

The Lucăceşti Sandstone Formation (pr), emerges among Bisericani layers, respectively the equivalent of Plopu Formation, at the base, and under Oligocene lower Menilite, which indicates the end of Eocene, both in the Vrancea unit and Doamna lithofacies. Because the thickness of these layers is decreasing to near extinction, within Bistriţa–Râşca subunit open in Vrancea Nappe, in the area of Cuejdiu hydrographic basin it emerges discontinuously, being mistaken by interstratified sandstones with brown marlstones and disodilic clays from the inferior member of Oligocene basal menilite (Atanasiu 1943; Ionesi 1982; Grasu et al. 1988).

For the petrographic description, due to the fact that the sequence from the analysed area is incomplete, information from the stratotype area identified in Tazlău digitation within the Tazlăul Sărat basin has been collected. Here, Lucăceşti Sandstone presents as a massive and hard rock combined with slightly consolidated sequences. The quartz arenite character and high petrographic maturity are tested by the high value of silica content (>90%) and low quantity of alumina (<3%) (Fig. 4.2).

The last two stratigraphical terms of the Palaeocene—Eocene sequence from Vrancea nappe, respectively Bisericani Formation and Lucăceşti Sandstone Formation, represent a particular interest to our study, because is considered that the landslide leading to the formation of natural dam lake Cuejdel through diluvial dam occurred on this lithologic background (Fig. 4.3).

A trigger factor of slope processes was represented by the low degree of compaction in the Bisericani Formation, consisting of a clay and argillaceous marl combination, in comparison to the higher toughness of Lucăceşti Sandstone, located in the upper part. Thus, the alternance of deposits with different toughness and compaction degree supported the emergence of landslide processes on the western slope of Muncelu Peak (Ionesi 1982; Grasu et al. 1988; Rădoane 2002).

Palaeocene–Eocene formations from Tarcău Nappe. Tarcău Sandstone Formation (Pg1-lt), is represented by an arenaceous flysch disposed in metric banks of 1–3, 3–5 and even over 10 m, among which red and green clay intercalations occur. Within the area of the Cuejdiu hydrographic basin, these emerge only in the upper sector, contributing to the lithologic support for Prislopaşu Peak (1213 m), the highest in the study area. The petrographic mosaic of Tarcău Sandstone indicates a large amount of graywacke and litho-feldspathic arenite, preceded by graywacke and

Fig. 4.2 Muncelu Peak (1177 m) and the main scarp of the landslide body (September 2016)

feldspatholothic arenite, respectively arenite and lithic graywacke. Thus, the characteristic features of Tarcău Sandstone are given by lithic wacke in brutish varieties and litho-feldspathic wacke in fine varieties (Grasu et al. 1988).

Oligocene formations from Vrancea Nappe. Lower Menilite Formation (if-ch), is stratigraphically continuing the analysed area that overlays Lucăceşti Sandstone and subordinates bituminous brown marlstones from the upper part, which indicates the start of the Oligocene, in the context of Kliwa lithofacies from the Vrancea unit area. Within the Cuejdiu hydrographic basin, in particular on the external flanks of the Doamna–Horaiţa anticlinorium, the formation presents two terms, one basal, mostly argillaceous–disodilic and a superior one of upper Menilite (Atanasiu 1943; Stoica 1944; Dumitrescu 1952; Ionesi 1971; Grasu et al. 1981).

Bituminous brown marlstones Formation (if-ch) occurs mostly in the upper sector of the Cuejdiu hydrographic basin, representing Pietricica lithofacies from the Bistriţa–Râşca subunit of Vrancea Nappe. The petrographic structure is characterized by sandstone intercalations disposed in banks of 0.5–1.5 m and conglomerates that occasionally tend to replace the Oligocene with a ruditic facies (Grasu et al. 1981). A specific aspect of the analysed area is also the development of the second level with brown marlstones in the lower disodile mass, also known as upper brown marlstones or interdisodilic brown marlstones. These emerge at a stratigraphical dis-

Fig. 4.3 Landslide headscarp view within the large sandstone boulders—Lucăceşti sandstone formation—pr, combined with slightly consolidated sequences of red and green clays deposits—Bisericani formation—pr (September 2016)

tance of 25–30 m, right under Kliwa Sandstone, disposed in batches of 2–4 m, 4–6 m and rarely of 10–15 m thickness (Băncilă 1958; Ionesi 1971; Ionesi and Grasu 1976).

Lower disodile Formation (if-ch) is generally characterized by different petrographic types, from which the bitumolite rocks have an economic importance. Within the Kliwa lithofacies of Tarcău Nappe, and also of Vrancea Nappe, disodile with a typical bitumolitic aspect hold the largest extension within the formation (Grasu et al. 1988).

Within Vrancea Nappe and in particular in the region of the Calu–Cuejdi–Cracău Valleys from the Bistriţa–Râșca subunit, the lower disodile get a peculiar aspect due to the emergence of green rocks buckets in their mass. The frequency and abundance of fragments determines, here and there, the replacement of initial disodilic deposits with genuine rudite having an argillaceous–disodilic matrix. Moreover, slightly bituminous disodile are not absent, having an arenaceous clay aspect that admit Kliwa Sandstone intercalations (Băncilă 1955; Mirăuţă 1962; Grasu et al. 1988).

Kliwa Sandstone Formation (if-ch) follows lower disodile and supports upper disodile and menilite, a specific stratification of extreme eastern regions of Vrancea Nappe, respectively in Bistriţa–Râșca subunit. Within the Kliwa lithofacies, these deposits represent an erosion limit (Filipescu 1936; Vinogradov et al. 1983). Along the Runc creek valley, the right tributary of Cuejdiu river, the deposits of this formation occur in heavy batches of 20–60 m, rarely of 100 m, sometimes being partially or totally replaced with rudite having green rocks elements, Dobruja type. Even so, typical Kliwa Sandstone also corresponds to some quartz arenites, both from a mineralogical and chemical point of view (Filipescu 1936; Mirăuţă and Mirăuţă 1964).

Upper disodile and menilite Formation (aq) is included between Kliwa Sandstone and Gura Șoimului Formation, representing a specific entity only for Vrancea Nappe, within which the formation sometimes ends the Oligo–Miocene sequence, when Miocene deposits are removed through erosion (Dumitrescu 1952; Ionesi 1971). Along Cuejdiu river valley, as in the case of lower disodile, the formation presents a large spectrum of petrographic types as: typical disodile, brown–black disodile, arenaceous disodile, disodilic clays and even greyish–green clays, marl disodile, black, bituminous and rarely even sideritic limestones.

Kliwa quartz arenites are developing in thin layers, in variable proportions, and in addition rubanate or blackish sandstones varieties can be mentioned. The menilites occur in the upper part of level, at first interstratified with disodile, then becoming compact, in batches of 12–16 m thickness, mostly micro-folded. An important characteristic of the last flysch unit is represented by the formation replacement through rudite with Dobruja green shale (Grasu et al. 1981).

Miocene formations from Vrancea Nappe. Gura Șoimului Formation (aq-bd) is presented as a petrographic term with a flysch character and partially molassic that continues the Oligocene sequence in Vrancea Nappe, belonging to early Aquitanian–Burdigalian. From a lithological standpoint, it represents a complex in which clays associate with arenites, siltstones and rudites, with menilites and disodiles characteristic to the Oligocene. Within the Cuejdiu hydrographic basin, the formation makes the transition from the monoclinal flysch area to the Subcarpathian molasse deposits, thus identifying itself mostly with the lithologic contact area of the two structural units (Stoica 1953; Ionesi and Gheaţă 1978; Ionesi and Bogatu 1986; Matei et al. 1986).

Pleistocene and Holocene formations. Pleistocene formations (qo ½ and qp 3/3) from the area of the Cuejdiu hydrographic basin are merged in the lower sector of the hydrographic course, outside the Carpathian area. They consist of diluvial–proluvial accumulations which have resulted mostly from loessoid deposits of interfluves or from old terraces deposits, upper or lower, containing sand, gravel and boulders, and also the remains of fossil mammals (Joja et al. 1968). Holocene accumulations (qh2) are represented by lower terraces and floodplain deposits, and also slope diluvial deposits, containing gravel and sand taken and transported towards the lower sectors of the main valleys.

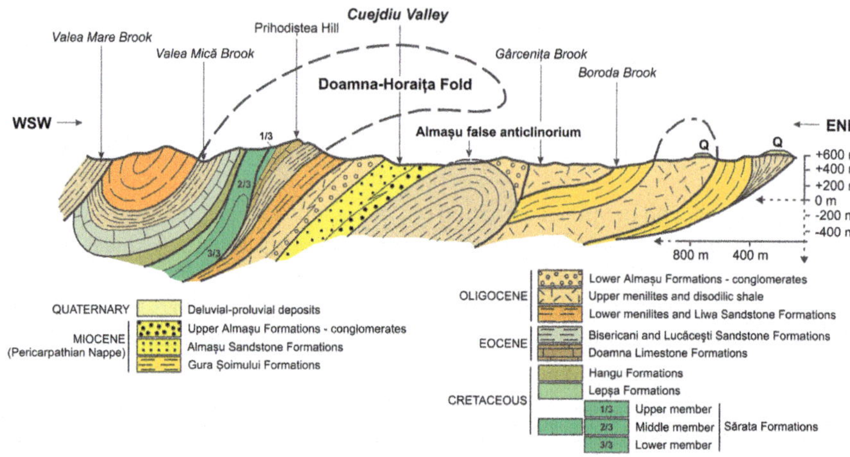

Fig. 4.4 A geological section of Doamna-Horaiţa Fold and Almaşu false anticlinorium from the middle sector of the Cuejdiu valley

4.1.2 Tectonic and Structural Aspects

The tectonics of the region are directly responsible for the complicated arrangement of the lithostratigraphical terms within Cuejdiu hydrographic basin, being induced by the presence of the Doamna–Horaiţa anticlinorium. In general, this represents the highest structure of the external flysch within the Bistriţa–Râşca subunit, consisting of a core in which a Cretaceous external facies of the flysch geosynclinorium is to be found located at the point of contact with Miocene molasse from Carpathian foreland. The upper part of this fold, formed from Palaeogene deposits, appears inverted towards east, this time over Miocene deposits, forming a structure named Almaşu false anticlinorium (Athanasiu et al. 1927; Atanasiu 1939, 1943; Mirăuţă 1962, 1969; Mirăuţă and Mirăuţă 1964) (Fig. 4.4).

One important lamination of Doamna–Horaiţa structure emerges in the reverse flank of the open fold in Cuejdiu Valley, where the lower horizon of Cretaceous comes in tectonic contact with Bisericani Formation. Northward from this sector, on Ţiganca creek valley, at a superior level of the fold, the fault continues bringing into tectonic contact the layers from Lepşa Formation and Late Eocene Formations.

Important laminations of Cretaceous and Eocene Formations also occur in the normal flank of Doamna–Horaiţa fold, where in Runcu and Tisa creek river beds and in Cuejdiu Valley, the Izvor Formation has only 20 m thickness, and only 10 m in early Eocene laminations. Also in the Cuejdiu Valley, between Varniţa creek and Cuejdi village, the axial area formations of the fold appear very narrow (200–500 m), represented by Cretaceous deposits with a diapir outcrop (Athanasiu 1905; Athanasiu et al. 1927; Atanasiu 1939, 1943; Mirăuţă 1962; Mirăuţă and Mirăuţă 1964).

Northward from the analysed area, in Cracău Valley, the tectonic contact between external flysch and Miocene molasse continues at the Oligocene deposits level with

Miocene—inferior deposits (the Oligocene deposits with Miocene—inferior ones), which is manifested through Oligocene horizons impairment and their thickness of only 100–150 m in Cuejdiu Valley.

Based on this, the fact was identified that the main thrust fault of flysch over the Pericarpathian foreland Miocene emerges on the territory of Cuejdiu river basin, just outward Almașu false anticlinorium. This aspect is emphasized on the right versant of Cuejdiu River, between Cuejdi and Gârcina settlements, and also between Slatina and Mitofan creeks, where conglomerates lift even up to 150 m above the river bed.

Oligocene deposits thrust, respectively menilite shale series, also continues from Gârcina to Piatra Neamț, where all over the eastern slope of Cozla Peak and under Pietricica, lifts in some places up to 60–70 m above the river bed (Mirăuţă 1962; Mirăuţă and Mirăuţă 1964).

Regarding the complex of faults from the western flank of the Doamna–Horaiţa fold, they have individualized a syncline with the faulted eastern flank (above Piatra Șoimului), where salt breccias emerge over Oligocene deposits. Southward, in the Runcu creek valley, the syncline with Oligocene deposits reoccurs due to an axial lift (Mirăuţă and Mirăuţă 1964; Grasu et al. 1988).

In the upper sector of the Cuejdiu hydrographic basin, west from the Doamna–Horaiţa fold, a parallel sequence of two anticlines emerges consisting of Eocene deposits well opened in Cuejdiu Valley, between the Recea and Gherman creeks. After quickly immersing southwards, these deposits reoccurred eastward from Bisericani Formation, on Vf. Mare Peak (1033.2 m). Between these anticlines two synclines are developing filled with Oligocene deposits, a western one flanked at the east by Gherman Peak (1053 m) and a south–eastern one, that crosscuts Oprești creek and continues to the upper sector of Runcu creek (Atanasiu 1939; Olteanu 1952, 1953; Mirăuţă 1962; Joja et al. 1968; Grasu et al. 1988).

4.1.3 The Paleogeomorphological Evolution

The paleogeomorphological evolution of the Cuejdiu hydrographic basin started at the same time with the amplitude lifting of the Doamna–Horaiţa fold and the overflow of folds across the Moldavian Plateau. Subsequent to this episode, a series of torrential organisms formed on the last peneplain before the formation of the actual hydrographic network, supported by the clay, marl and arenaceous–marl lithologic facies. From that moment a new stage of morphogenesis commenced, named the neogeomorphological subphase, in which the Cujediu Valley was probably initiated, yet having a lower extinction that the initial shape (Ionesi 1971; Ionesi and Grasu 1976; Ichim 1979).

In the initial surfaces and Sarmatic piedmont phase, similar to many homonymous rivers in the area (Suha Mare, Suha Mică, Ozana, Cracău, Almaş), Cujediu was well developed in the relief, yet contrasting its short length with its age. This fact indicates a development on a relief surface with low energy and inclined towards Sarmatic Sea, which directed the initial course more eastwards from the actual direction, towards Șerbești Hill and Mărgineni Peak area.

After this phase, the valley developed until the Doamna–Horaiţa tectonic lift, adapting its course to the western morphotectonic contact, becoming a tributary of Cracău at the 150 m terrace, and at the 40 m terrace a left tributary of Bistrița, after a torrent catchment from the south versant of Cozla–Pietricica Peaks. Around the same time, were found some network reshuffles supported by lithologic permissiveness, similar to the catchments of all tributaries in the upper basin (Țiganca Brook, Cuejdel Brook, Pârâul Argintăriei Brook, Cerbu Brook, Vaca Babei Brook, Bursucăria Brook, Mestecăniş Brook), or meander catchment from Cuejdiu Valley, upstream of confluence with Runcu creek (Polonic and Polonic 1968; Ionesi 1971; Ionesi and Grasu 1976; Ichim 1979).

The present shape of valleys certifies a more pronounced influence of tectonics than lithology, especially in the middle and lower sector where the main course follows the contact between flysch unit and Subcarpathian unit, generating a pronounced asymmetry of height between the two river banks. The slightly dendritic aspect from the upper sector of Cuejdiu highlights a low influence of geologic structure consisting of lithologic alternations mostly permissive and even soft (Ichim 1969, 1973, 1979).

4.2 The Role of Geomorphological Factors

In order to geomorphologically analyse the Cuejdiu basin, a geospatial database was created in GIS (Geographic Information System). This was done using the software dedicated to spatial analyses: TNTmips, ArcGIS (ArcMap; ArcScene; ArcCatalog) and Global Mapper. As cartographic support use was made of: topographic plans (scale 1:5000; 1985 edition); forest plans (scale 1:10,000; 1981 edition); topographic maps (scale 1:25,000; 1984 edition); ortophotoplans taken during the flights of 2005 and 2012, as well as LANDSAT satellite images. The database was completed with direct topographic measurements, using the Leica 1200 System, made of Leica TCR 1201 Total Station and Leica GPS 1200, although the high declivity of relief and the high degree of afforestation made this more difficult.

The digital elevation model (DEM) was obtained after following certain intermediate stages: collection and cartographic fund selection; import in GIS software and georeferencing the cartographic objects in the national projection system of Romania (STEREO 70); manual vectorization of isohypses (equidistance 2.5 m and 5 m—scale 1:5,000; 10 m—scale 1:25,000) and assigning altitude values; taking altitudinal quotas and hypsometric quotas apposite to the hydrographic network indicated on cartographic support (>1000 altitudinal quotas), and supplementing them with the GPS database. After interpolation of the hypsometric curves, altitudinal quotas and morphometric information induced by hydrographic network, by Topo to Raster command (ArcGIS 10.1) a detailed model of the topographical surface from the analysed area (DEM rezolution—1 m^2/pixel) was obtained (Fig. 4.5).

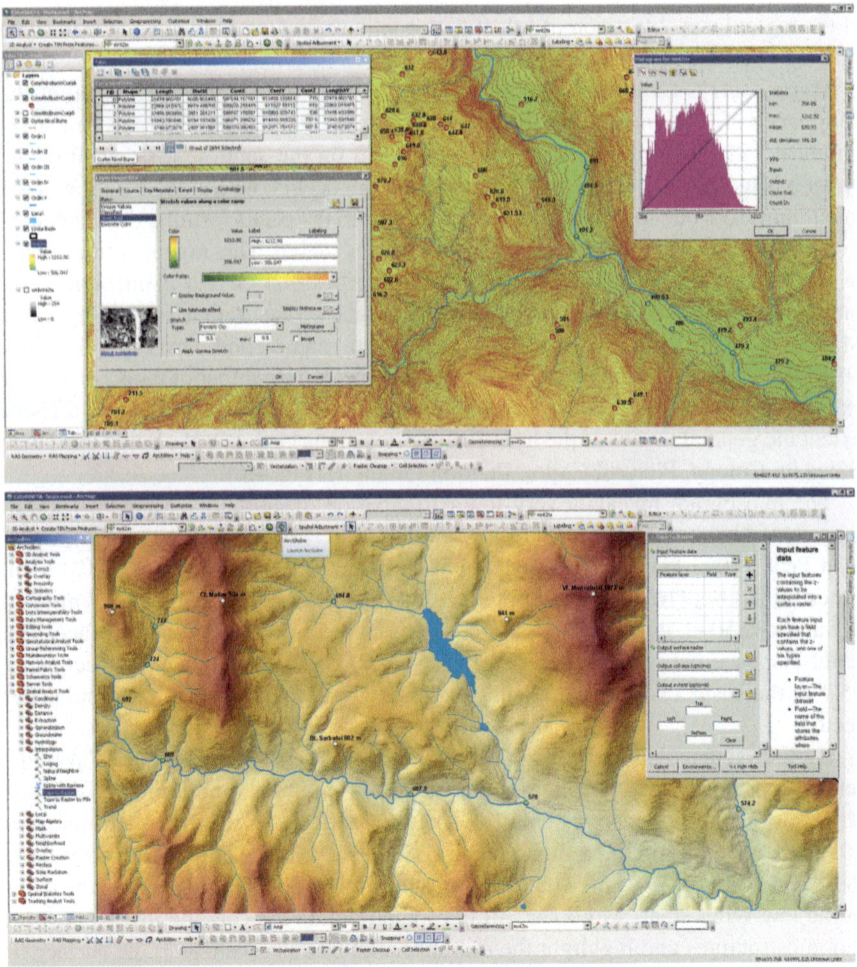

Fig. 4.5 The working method in ArcGIS 10.2 created for the geospatial database

4.2.1 Morphographic and Morphometric Issues

The relief of the Cuejdiu River basin corresponds to the Eastern Carpathians general context in which it has evolved, its morphometric and morphographic peculiarities being highlighted by the morphological contact area between the southeastern part of the Stânişoarei Mountains (subunit of Neamţ Mountains) and the Pericarpathian region. The main features of the relief can be summarized as follows: well-expressed structural relief elements along the lines of tectonic subduction; relief heights located generally below the ceiling of 1000 m; relief energies generally below 400 m for most of the region; a proliferation of selective modeling relief, and the main role in slope shaping due to landslides (Ichim 1969, 1979).

Hypsometry. Elevations within the basin ranges from a low of 306 m, at the confluence of Cuejdiu and Bistriţa in Piatra Neamţ (Dărmăneşti district), to a maximum of 1213 m in Prislopaşu Peak, positioned on the southwest sidewall of the upper basin. Thus a maximum difference of 907 m occurs, with a depreciation of the maximum from outside to inside, respectively from the peaks forming the watershed to the main valley, where the lowest values are registered. This layout is interrupted in the northern basin just by the transverse interposition of two main ridges, virtually delimiting the Cuejdel creek basin and the lake basin, first by the alignment Măgura Hill (1117 m)—Muncelului Peak (1077 m) and secondly by Tarniţelor Peak (1081 m)—Mătieş Summit (986 m). To these are added a number of flanks or shorter extensions of the main peaks descending into the basin, but disturbing to a lesser extent the general layout of the hypsometric steps (Burlacilor Peak, Bâtca Pârliturii, Bâtca lui Coman, Piciorul Cailor, Piciorul Draniţei) (Fig. 4.6).

The hypsogram reveals a very low participation of altitudinal values >1000 m represented by a series of peaks, ridges and summits: Vârful Mare (1033.2 m), Gherman Peak (1053 m), Vaca Babei Hillock (1104.5 m), Prislopaşu Peak (1213 m), Bâtca lui Andrei (1079.9 m), Tarniţelor Peak (1081 m), Măgura Hill (1117 m), and of values <350 m, corresponding to the lower basin outside the folded surface. The relief disposed between 650 and 850 m reaches the highest frequency, occupying 35% of the basin, characterizing almost all altitudinal steps of the middle and upper sector.

By the rectangular-grid method (1 square = 1 km^2) for an area of 97.49 km^2 related to Cuejdiu River basin, average values of maximum, minimum and average altitude of the region were calculated. The average value of maximum altitude was 809.5 m, and of minimum altitude was 611.8 m. The average altitude of the basin in these conditions is 704.5 m, which confirms the morphological evolution of the relief in the low-lying areas. Moreover, all hypsometric data indicate the involvement of morphostructures and highlight the morphosculpture role in shaping the surface. It was concluded that actual relief is young and evolving.

Density of the hydrographic network. By analysing the map of relief fragmentation based on the density of the river network, it results that the lowest values (<1 km/km^2) correspond to the drainage divide of the main basin and to the relatively narrow interfluves of the secondary river basins, where the runoff does not qualify for forming the main hydrographic network. The highest values recorded are over 2–2.5 km/km^2 and correspond to certain sectors along the main course, with greater expansion on the right bank of the upper basin, and partly on the lower left sector where, because of the short length of the slopes torrential organisms are installed that fragment the landscape. The same phenomenon is encountered in the basin of Runcu, Cerbu, Cuejdel or Gârceniţa streams, because morphology does not allow the development of more significant catchment areas. In both cases, in registering these values are included the main beds, the secondary beds and the torrential formations.

Within the natural dam lake Cuejdel area, either we talk about the catchment basin, or about the downstream slopes, the density of relief fragmentation averages

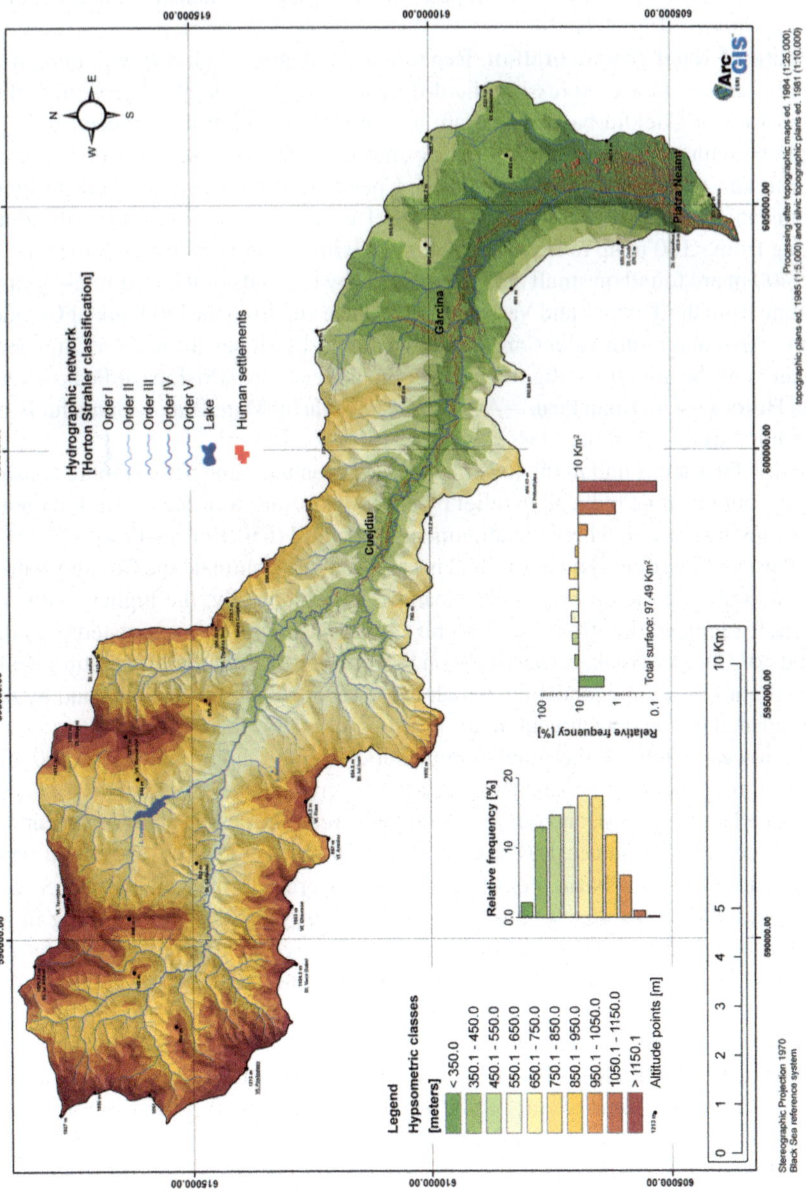

Fig. 4.6 Hypsometric map of the Cuejdiu hydrographic basin (DEM rezolution—1 m^2/pixel)

around 1–2 km/km^2. It is worth noting the low value of this indicator on the Făgetul creek valley, left tributary of the lake and on the Piciorul Crucii valley, right tributary, where riverbeds are deep incised in relief, but without major tributaries. This induces a compact character of the lake slopes, the river network density not exceeding 0.8–1.2 km/km^2 (Fig. 4.7).

Depth of relief fragmentation. Regarding the depth of relief fragmentation or the relief energy index expressing the different penetration depth of erosion within the territory of Cuejdiu basin, there are a number of peculiarities induced both by morphostructure and by the hypsometric index which decrease from upstream to downstream, respectively from the exterior peaks to the mainstream bed. In these circumstances it was observed that the relief energy has a wide range of values ranging from <100 m up to less than 400 m. Maximum values between 350 and 400 and >400 m are found on small areas dominated by the folds of the Doamna–Horaiţa anticline from the Ţiganca and Varniţa creek basins, and from the left flank of Cuejdel stream. Also maximum values are induced in relief by slopes formed on Oligocene structures of the anticlinal alignment from the upper basin (Prislopaşu Peak—Vaca Babei Hillock—Gherman Peak—Arinilor Peak—Vârful Mare Peak—Bâtca lui Ivan; Bâtca lui Andrei—Gârcina Massif—Mătieş ridge).

High values are found in the lower basin, based on the same petrographic considerations, but this time induced in relief by the contact zone between the mountainous and the lowland areas, where the alignment Cozla Hill (650.19 m)—Gospodina Hill (419.9 m) and Pietricica Ridge (527.82 m) altitudinally dominate the Cuejdiu Valley at the entrance to the city of Piatra Neamţ. Not incidentally, the highest value of the relief fragmentation from the Cuejdiu stream basin (430 m) is registered on the left side of Cuejdel creek, respectively on the western flank of the Muncelului Peak, induced both by a large landslide, which led to the formation of the lake and by the slope formed by the detachment area.

The lowest values of the relief fragmentation depth ranging between <100 and 100–150 m occur in some sectors of the Runcu creek basin, in the central part of the Cuejdel stream basin, and outside the folded area. The largest areas are occupied by the categories of values 150–200 m, 200–250 m, and 250–300 m, which have amounted >50% of the basin. For that matter, the average of maximum relief energy is 242.4 m and the average of minimum relief energy is 200.1 m, resulting in an average power for the entire basin of 221.2 m. As far as the entire analysed surface is concerned, these values describe a high homogeneity of the morphometric indicator (Fig. 4.8).

Declivity. Another important parameter in describing geomorphological factors installed in the Cuejdiu River basin is land declivity. It lies in the relationship between relief fragmentation density and depth, petrographic composition and tectonic configuration, which cumulatively print differentiated dynamics of the modeling processes. In the analyzed area there are great variations of slopes. They start from a few degrees along the valley corridors (<5°), due to asymmetries from the lower basin and from the left side of Cuejdiu, reaching values in excess of 50°–60° generated by the main peaks (Mătieş ridge, Gârcina Massif, Prislopaşului Peak, Muncelului Peak, Vârful Mare Peak, Arinilor Peak, Pietricica ridge etc.). High gradient values are also

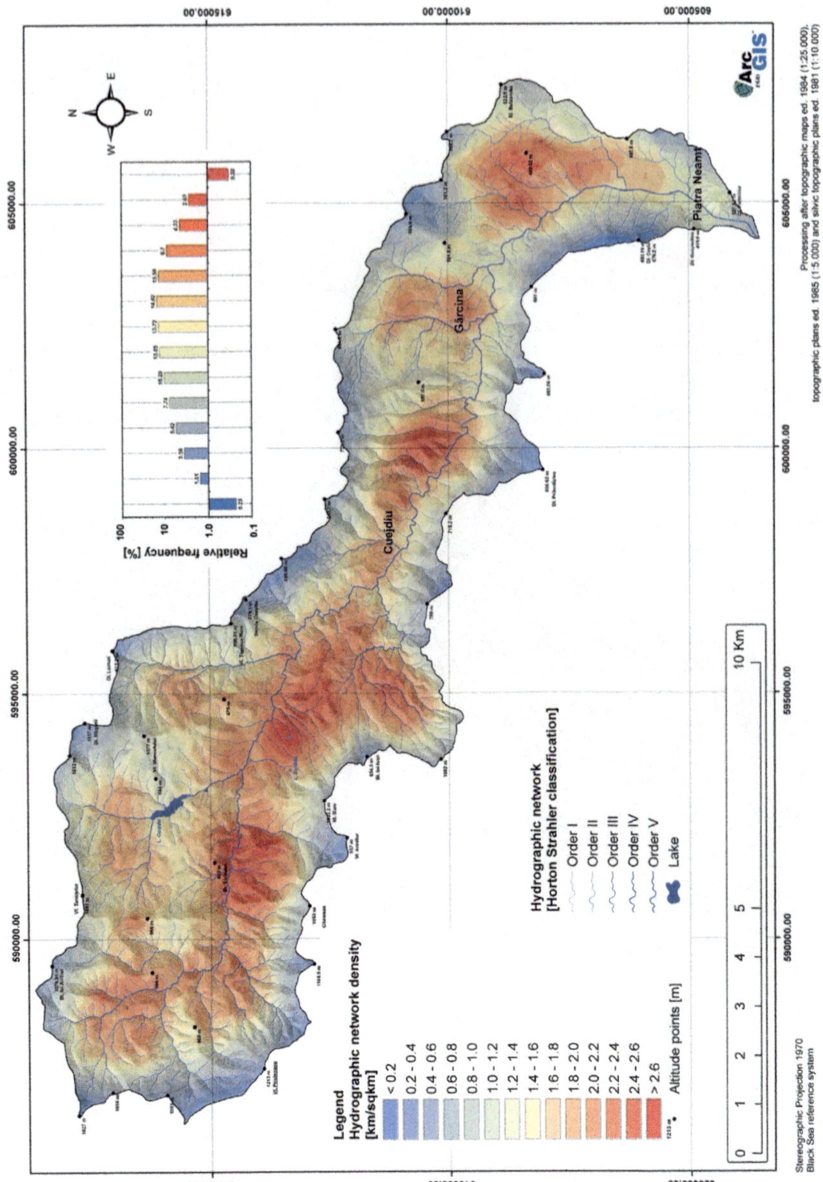

Fig. 4.7 Density of the hydrographic network map of the Cuejdiu hydrographic basin

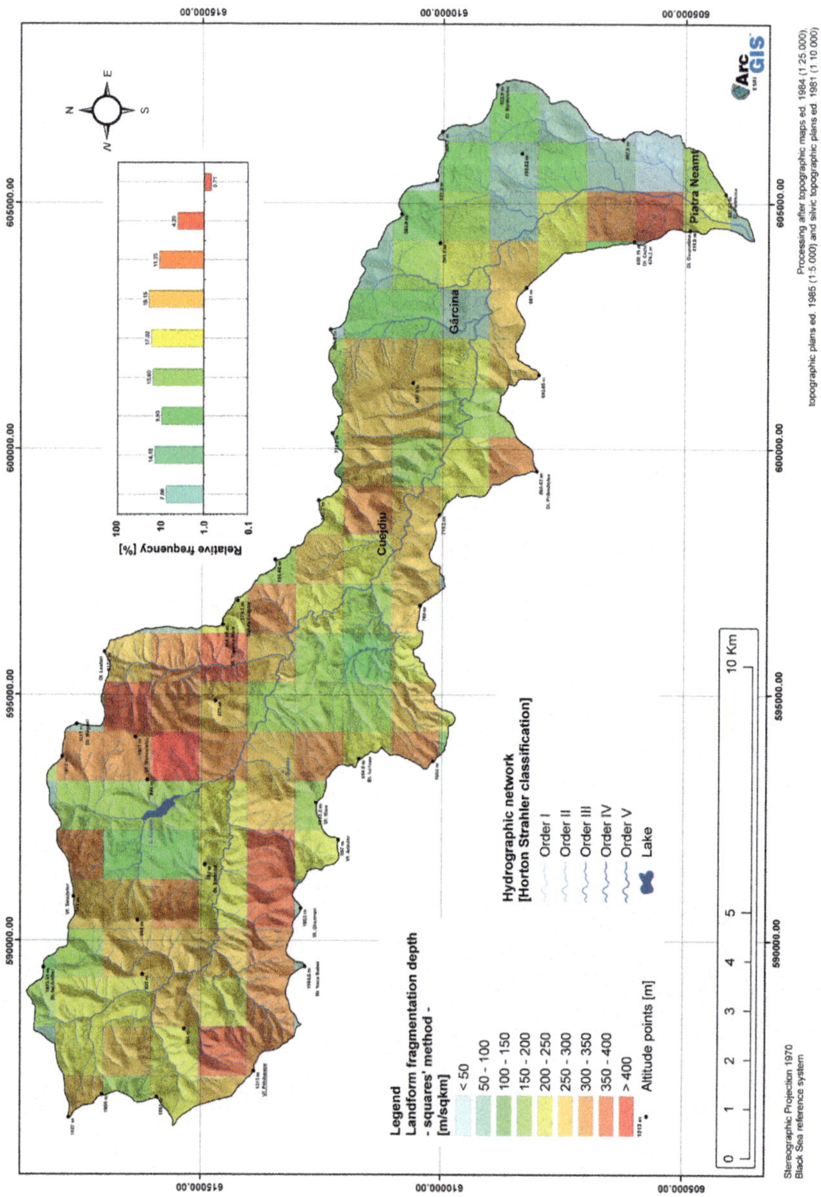

Fig. 4.8 Depth of relief fragmentation map of the Cuejdiu hydrographic basin

recorded in sectors of lithological erosion of the main watercourses: on Cuejdiu River (Cuejdiu locality sector, the sector under Sârbului Hillock, Piciorul Rotund sector), or on Ţiganca, Argintăriei, Cerbul, Prislopaşu, Strigoieşu, Pascău and Vaca Babei creeks valley.

Areas with declivity between 15° and 25° are well represented throughout the basin area (33.62%) and make the transition between the valley corridors and the main peaks. Range of 25°–35° generally characterizes the declivity of all slopes generated by the hydrographic network formed within the folded area. In the background of the same range the large-scale landslide occurred that resulted in the formation of the Cuejdel Lake (Fig. 4.9).

Slope aspect. An important role in describing the geomorphological factors is represented by slope aspect, because their arrangement affects the distribution and evolution of hydro-climatic phenomena as in the case of rainfall spatial distribution, the general direction of the wind, etc. Within the examined basin it is noted that the right side is almost entirely dominated by North, Northeast and East orientation, and the left side is dominated by the Southern, Southwestern and mainly Western orientation. The same layout is largely followed by the upper basin, where on the left side of Cuejdiu, the streams Cerbul, Cuejdel and Ţiganca redirect the general slope aspect, being characterized by left slopes oriented towards the West, and the right slopes consisting of a mosaic dominated by Southern, Eastern and Northern orientation.

The layout of the slopes in this order, and the relative homogeneity of the entire basin balance are induced by the general direction of the main valley, which flows along the Northwest–Southeast alignment (Fig. 4.10)

4.2.2 Landforms and Genetic Types

Geomorphological peculiarities encountered in the Cuejdiu river basin correspond to the morphostructural unit of Bistriţa, dominated by the morphotectonic factor. For this reason, the variety of genetic types and landforms is subordinated to modeling processes, differing according to lithological structures with different hardness and disposal, brought in atypical alternation of endogenous factors.

Petrographic and structural relief. Layout in overthrust sheets requires the analysis of morphostructures in terms of elements highlighted by internal tectonics due to the presence of small nappe-type folds, of faults and digitation, or of tectonic units and subunits materialized on the surface in the complex lithological conditions of flysch. In this light, in his study on the Stânişoarei Mountains, Ichim (1979) identifies three petrographic complexes within the area of the Cuejdiu Valley with high, middle and intermediate resistance to modeler factors, and a complex characterized by resistance to modeling due to lithological heterogeneity.

The lithological complex with high resistance to modeling is induced in relief by the presence in the structure of crests from Tarcău Sandstone, Kliwa Sandstone, Lucăceşti Sandstone or Jgheabu Mare Formations. Representative from this point

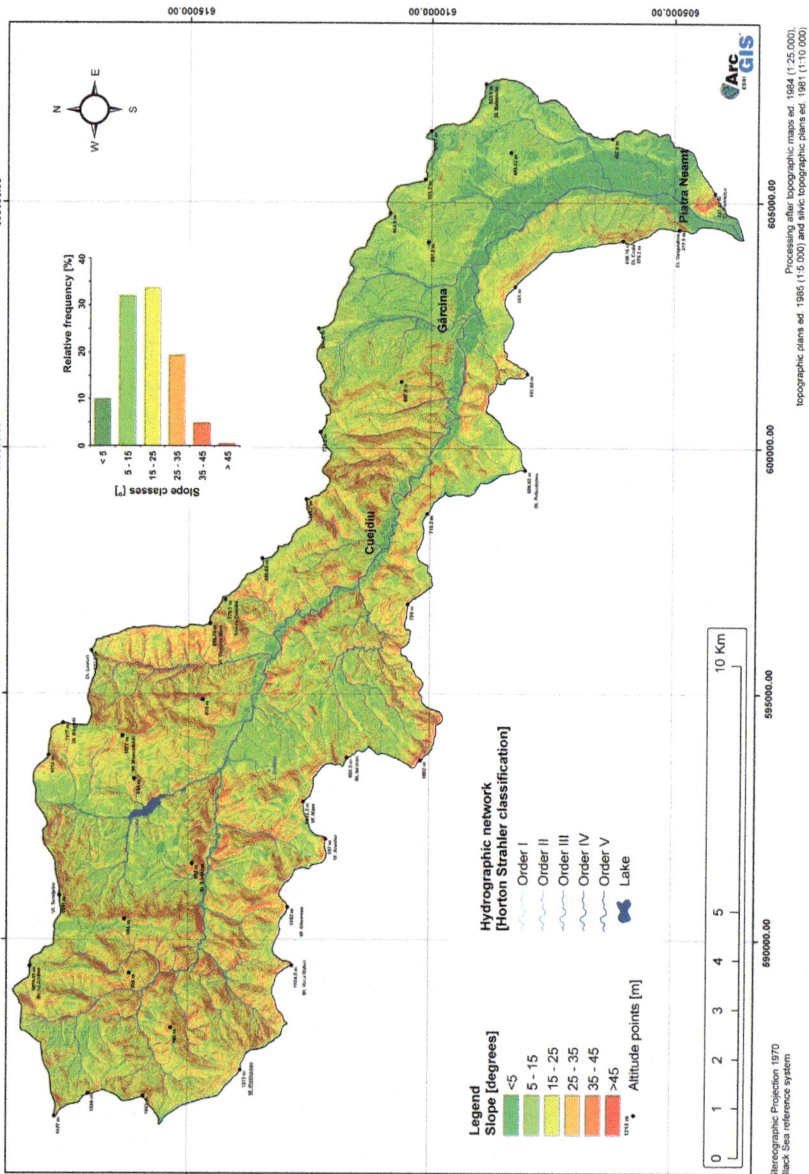

Fig. 4.9 Slope map of the Cuejdiu hydrographic basin

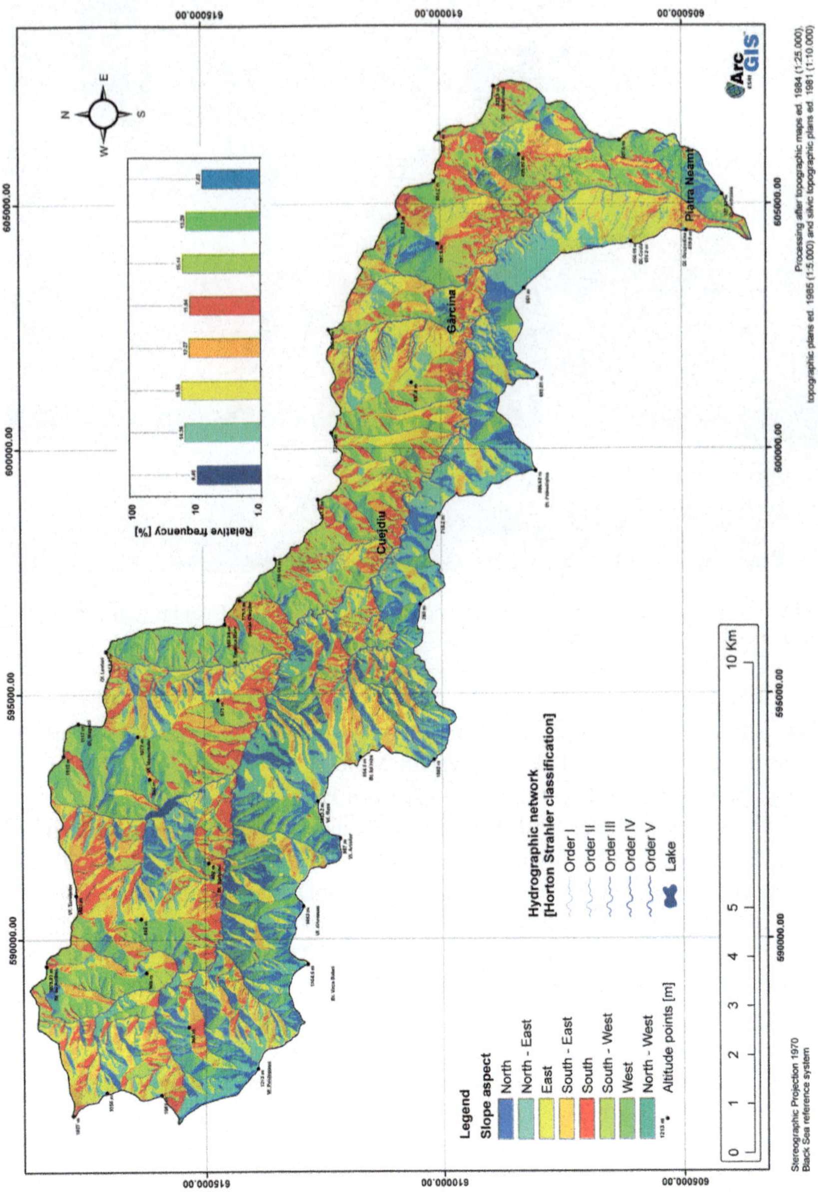

Fig. 4.10 Slope aspect map of the Cuejdiu hydrographic basin

Fig. 4.11 The lithological complex with high heterogeneity opened in Cuejdiu valley by erosion (September 2016)

of view is a series of interfluvial monts or peaks induced in relief by the highest altitudes: Prislopaşu Peak (1213 m), Vârful Mare Peak (1053.2 m), Măgurii Hill (1117 m), Muncelului Hilltop (1077 m), Pârliturii Hillock (966 m).

The lithological complex with medium resistance is due to the presence in the structure of crests of Audia, Izvor (Runcu) and Piatra Uscată Formations, and when they hold considerable thicknesses, to the presence of Doamna Limestones. The most representative monts formed from this lithostratigraphical alternation are: Ţiganca Mare Peak (956.8 m) and Prihodişte Hill (806.6 m). The lithological complex with low resistance is formed in the mountain area from the alternation of Hangu, Lepşa and Bisericani layers (Mătieş Mont 986 m, Sârbului Hillock 802 m, Sonda Cuejdiului Hill 779.1 m), and in Pericarpathian area from the presence of Miocene deposits, such as those from the sequence of Gura Șoimului Formation (Mintiana Hill 594.2 m, Balaurului Hill 522.9 m). The lithological complex with high heterogeneity is due to petrographic alternation in small areas, without significant rhythmic sequence induced by the presence of Paleogene flysch: Bâtca lui Andrei (1079.9 m), Vaca Babei Hillock (1104.5 m), Gherman Peak (1053 m), Tarniţelor Peak (1081 m), Cozla Hill (676.2 m), Pietricica Hill (527.8 m), Gospodina Hill (419.9 m) (Fig. 4.11).

Within the same petrographic complexes there is also highlighted, in the actual relief, the report of the valleys with their structure and lithology, which determines differentiated situations, generating types of longitudinal, diagonal, transverse and tectonic contact valleys. Based on the order of size of the hydrographic network (Horton-Strahler classification) that describes the valley system from the analysed basin, a map was generated of the relationship between the direction of watercourses and the general strike line. The analysis used as representation sequences the river sectors located between two confluences (Fig. 4.12).

As for the main stream (5th Order), it is noticed that the Cuejdiu Valley is dominated by sectors situated at tectonic contacts (38.78%) which set the valley direction on the faulted structures of Doamna–Horaiţa anticline or on the alignment of flysch folds discharge over the Pericarpathian area. In the upper sector of the basin, the valley runs mainly diagonally and even perpendicularly on the general strike line direction (>50%). This layout emphasizes that lithology is permissive, the region tectonics becoming the main argument for directions or asymmetry, and for the emergence of lithological thresholds along the main valley.

Concerning the lithological relation of the 4th Order valleys, those situated on the left of the main stream (Gârceniţa, Ţiganca, Cuejdel and Cerbu), have longitudinal disposal, emerging diagonal or transverse tributaries, and those situated on the right (Prislopaşu, Vaca Babei and Runcu) are mainly diagonally arranged, taking tributaries with chaotic disposal. In this way a series of structural scarps ranging between 100 and 300 m and dominating the right bank of Cuejdiu, is revealed downstream of Pârâul Adânc to the confluence with the Runcu creek, and in the sector next to the school from Cuejdiu settlement.

Lithologic slopes caused by erosion on layer extremities arise in the case of 4th Order valleys draining the right side of the basin, and in the valleys of inferior rank (3rd and 2nd Orders), tributaries to the 4th Order valleys from the left of Cuejdiu. In the case of the 1st Order segments, the relatively low depths of the valleys, and the reduced influence of lithology indicate a weak correlation, which result in an equal proportion between longitudinal, diagonal and transversal layouts, and a small share of those situated on tectonic contact. In the case of the Cuejdel Valley, scarps are present on the left slope of Cuejdel Lake (the west slope of Muncelului Peak) and downstream, as a consequence of the diagonal layout of the valley on the lithologic structures. This layout was also the gravitational support of the landslide which obstructed the valley. Up the lake, the 3rd and 2nd Order tributaries (Glodu and Făgetu) transect the slopes due to their almost transversal layout, generating lithologic slopes along the riverbeds.

Periglacial relief. During the Pleistocene, the southeastern part of Stânişoarei Mountains was found in the area of mountain permafrost, the periglacial landforms being materialized by the presence of solifluction deposits highlighted on the terraces and on the alluvial cones. In the Cuejdiu Valley area they appear in the structure of alluvial fans of Sălatruc and Cuejdiu Streams at the absolute altitude of 525 m (Ichim 1969, 1979). Rubbles occupy lower areas compared to neighboring mountainous regions, and are related in the Cuejdiu basin to the island occurrence of Kliwa and Tarcău hard sandstones or to the Izvor (Runcu) layers. The lower limit is maintained at an altitude of 650–850 m, where they appear inside the deluvial sliding mass under Muncelului Peak (1077 m). Even if they are mostly imbedded by forest vegetation,

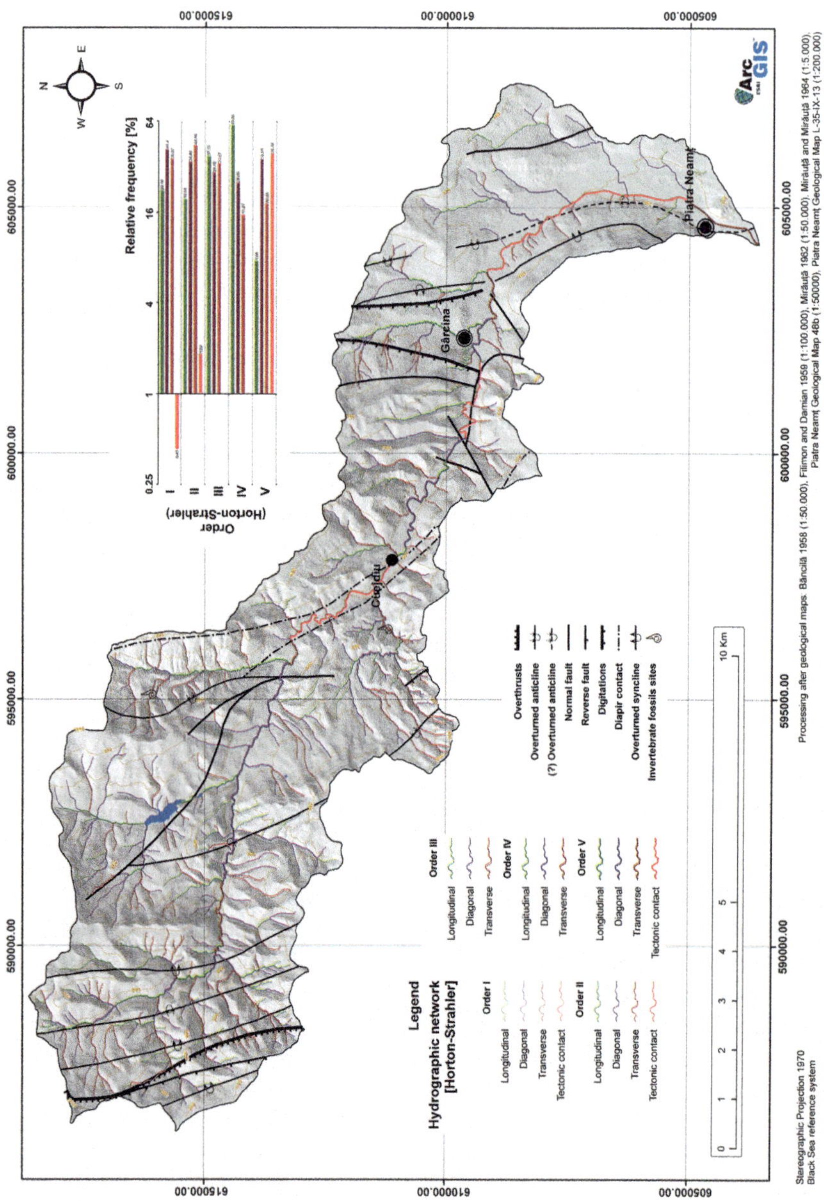

Fig. 4.12 The map of valleys report with their structure and lithology on the Cuejdiu hydrographic basin

the gelifracts brought to the surface attest, based on their polyhedral shape and size between 10 and 20 cm, the presence of periglacial alteration deposits in the upper basin Cuejdel.

Fluvio-denudational relief. Within these genetic types of relief, it is highlighted the network of primary valleys that are the result of a long evolution, marked by a series of catchments of the tributaries of the upper basin of Cuejdiu and a meander catchment upstream the confluence with the Runcu creek. Early sculptural forms are represented by gutters and gullies that appear mainly in the lower basin, where more recent deposits and low vegetation cover allow the initiation of torrential erosion. Ravines and cliffs are grouped furthermore in the mountains, where linear erosion is accelerated by high declivity and the presence of both permissive lithological formations and their derivative deposits.

The primary 1st and 2nd Order valleys are characterized by a lack of proper minor beds and floodplains, and by the absence of river terraces that appear only in the form of alluvial cones in confluence areas. Another important feature is their symmetry induced by the intersection of slope bases along the thalweg. Furthermore, basic valley benches are comparable to the rock terraces that occur in the transverse valleys. Even if the role of structure and lithology does not motivate their existence, it is considered that these benches can only be pseudo-river terraces. These oscillations of the local basic level occur in almost all valleys of inferior rank (2nd–3rd) upstream of the confluence of Cuejdiu and Ţiganca creeks (Ichim 1979).

Accumulation forms usually have a colluvio-proluvial nature, such glacis formed at the contact between the slopes and terraces higher than 4–6 m. Their maximum length is 150–200 m, their thickness is variable, but mostly reduced, and their age is Holocene, for those formed at first terraces linkage, and Würmian for those from terraces higher than 4–5 m.

Fluvial relief. In this category were included both the relief forms generated by the permanent network through erosion, respectively active banks and riverbeds, and those formed by the accumulation processes of fluvial terraces type. The active banks evolve due to lateral erosion, which makes the primary valleys (1st and 2nd Order) develop only through the stationary rock, as is the case of the Făgetu Stream, the left tributary of Cuejdel Lake. In the case of higher rank rivers, they develop both through the stationary rock, and through their own alluvia, the two types alternating until the exit of Cuejdiu from the mountain area, where the active banks manifest only through the alluvial deposits (Ichim 1979) (Fig. 4.13).

In the case of 3rd Order valleys, these have the beds sunken in silt and prolluvia along almost all their course (Carpenul, Pârâul Opreşti, Părul, Mesteacănul, Bursucăria, Pascău, Strigoieşul, Pârâul Adânc, Argintăriei, Muncelu, Varniţa and Sălătruc streams). Valleys of 4th and 5th Order present, with few exceptions, depths between 0.5 and 1.5 m, and in areas where the bed is steeped in stationary rock, frequently exceeding depths of 3–4 m. In both cases, minor riverbeds are permanently covered with a layer of silt at least 20–40 cm thick. Also, in the middle and lower course of Cuejdiu, many sectors of riverbed with braided streams appear. For this reason, the width of minor beds varies from 8–12 m to 30–40 m.

Fig. 4.13 Fluvial landforms in upper Cuejdiu hydrographic basin (September 2016)

Regarding the terraces level from the Cuejdiu River basin, in the mountainous area terraces were identified up to the relative altitude of 35–40 m, and at the contact with Subcarpathian area, at a relative altitude of 160 m. Thus, in the middle sector of the basin are four, five and even six levels, and the number of terraces increases to nine in the area of molasse deposits. Their level aims all along Cuejdiu river almost the same amplitude and can be identified among the following values: 2–4 m, 5–7 m, 9–12 m, 15–20 m, 22–25 m, 35–40 m, 85–90 m, 100–120 m and 140–160 m, the last corresponding to the interfluvial peaks Mintiana and Balaurului (Ichim 1979) (Fig. 4.13).

The lowest terraces occupy small areas and consist of mixed deposits alternating sometimes with alluvial deposits due to irregularities of the riverbed sunken in alluvium. The 9–12 m terrace is best conserved in relief, and its mixed character makes the alluvial thickness move from 0.5 to 1 m in the mountainous area to full alluvial in the Subcarpathian area. The other terraces appear fragmented, especially in the lower basin sector, where sediments carried by the river are becoming increasingly important.

Anthropic relief. The Cuejdiu Valley is a well populated region in its lower half, where the settlement density increases with the river exit from the mountainous area, respectively from Cuejdiu and Gârcina settlements towards Piatra Neamţ. Changes in the physical-geographical landscape are a consequence of natural potential harness, but due to human intervention that took place progressively, today is registered a rebalancing of new environmental conditions. Among the most important anthropogenic landforms we can signal a number of facilities in the first three terraces of Cuejdiu (2–4 m, 5–7 m, 9–12 m), due to construction location, to modification and impoundment of active banks, to channeling of the minor bed in the urban area, etc.

In the upper basin of Cuejdiu or in the natural dam lake Cuejdel, anthropogenic intervention materialized by building forest roads on the slopes. This was also a powerful factor for destabilizing the slopes under Muncelului Peak, breaking their gravitational balance. After this episode, an anti-erosion barrage was made at the lower part of the deluvial mass in order to stabilize materials accumulated on the Cuejdel stream valley. Thus, anthropogenic landforms in this sector are reduced only to a transport network that pursues the main valley and a few minor interventions at the level of riverbeds, natural landscape still preserving an unaltered form.

4.2.3 Present Geomorphological Processes

Given the evolution of the Cuejdiu hydrographic basin in the Carpathian flysch area, a region that is characterized by a very wide range of geomorphological processes and phenomena, it is evident that of all the relief shaping agents, water erosion holds the most important part. It may lie at the core of process triggering conditions in permissive lithology, but it can also transform, through the draining network, into a direct factor for modeling and transportation. In both cases, water and its dynamics control the present geomorphological processes that occur in the reference area. Out of these processes, the mass displacement ones are best represented, vigorously dominating versant morphodynamics (Ichim 1973).

Based on conducted research, six types of gravitational processes that affect the south-eastern region of the Stânișoarei Mountains were identified: failures, rolling, creeping, landslides, solifluxion and mud-flow. In the Cuejdiu hydrographic basin we will only limit our research to the processes that best come forth in the landscape, respectively landslides, solifluxion and creep.

Landslides. The largest areas affected by landslides on the Cuejdiu Valley are found in the mountainous sector, on the high slopes and with a complex lithology. Because of the high foresting degree of the middle and upper section of the basin, which confirms the considerable antiquity of the landslides, most probably Pleistocene, mapping is hindered a lot. Still, the most important gravitational displacements have been recorded in the area of the Runcu, Părul (Ponoare Hill, junction with Cuejdiu River), Cuejdel and Sălătruc creeks, as well as on the eastern side of the Cozla Peak. A series of superficial slides are also noticeable at the contact with the Subcarpathian area, in the upper sector of the Gârcenița creek or on the western and southern versant of the Mintiana Hill (Fig. 4.14).

As most landslides occurred at the beginning of the Quaternary, in the present phase we can only register microrelief shapes dwelling in the landscape as a testimony of ancient versant processes. At the present moment, only a range of simple shapes

Fig. 4.14 Present landslide processes in the Cuejdiu hydrographic basin (September 2016)

can be spotted, such as scarps of old landslides, stabilized thanks to their immersing in the blankets initially mobilized. Out of these, only the more recent cornices are active, most of them formed on old diluvia, and they are characterized by smaller surfaces and low amplitude, not higher than 5–7 m. The microrelief formed on the landslide waste is also worthy of note, as it is made up of the monticles, catsteps and slip-wrinkles as well as a combination of these, a process generating more complex shapes.

Mound-shaped microrelief presents a large height variation relative to the slopes record, reaching 15–20 m. Behind them, a series of intermound microdepressions are found that often supply the frame for genuine lacustrine basins. In the Cuejdiu basin, the formation of such a lake is signaled on the western side, in the place known as "La Ponoare", which in the year 1979 was 3500 m^2 and 9.5 m deep. At the present time, the lake known under the hydronym Bolătău has reduced its depth with -1.5 to -2 m, given the pronounced clogging and in situ biodegradation through forest vegetation accretion (Ichim 1979; Mihu-Pintilie et al. 2014).

Catstep or landslide terrace microrelief does not have the same spread of the monticular type, but it does appear in the Cuejdiu Valley along the main course, at the relative altitude of 20–30 m. In terms of morphometry, the landslide benches hold a length of 150–200 m, relative dislevelments of up to 30–40 m and the width of the terraces varies between 7 and 30–40 m. In the case of slip-ridge microrelief, their height reaches up to 10–15 m and their length varies between 70 and 90 m. They are best highlighted by the presence of talus caught in the slip ridges on the eastern versant of the Cozla Peak.

In the present phase, landslides still hold an important part in modeling the versants and are, most often, a consequence of old landslides revival. Excess humidity that characterized the 1970–1991 interval led to versant processes restart, with a considerable spread at least in two sectors of the hydrographic basin of Cuejdiu. The first event took place in 1971, when in the Runc brook basin, an old landslide reactivated and mobilized a 3–4 m thick diluvium.

The second event which, for that matter, led to the formation of the Cuejdel Lake, had as a starting point a landslide that disrupted the rock in situ, covering a 35 ha surface and took place in two stages, on the left slope of the Cuejdiu Valley. The first gravitational dislocations started in 1978, when the diluvial mass dammed the valley for a short time. Later, in the year 1991, the diluvium reactivated on a much bigger scale impounded the whole watercourse at 0.5–1.5 km upstream from the confluence with Cuejdiu, determining the formation of the biggest natural dam lake in the Eastern Carpathians.

The main triggering condition was attributed to abundant rainfall in the interval of May–August 1991, when the added value was 741.4 mm. At the present moment, the deposition layer varies between 5 and 25 m or even more in the area of the dyke, and their evolution is still in a relative balance. The slope gravitational processes might reactivate due to the fact that in 2003–2004 a new forest road was built, connecting Magazia (Crăcăoani) and the left shore of the Cuejdel Lake (Ichim et al. 1996; Rădoane 2002; Mihu-Pintilie and Romanescu 2011; Mihu-Pintilie et al. 2012).

Solifluxion. Solifluxion is a geomorphological process similar to superficial landslides, combined with flows that carry along the prevailingly fine diluvia with high plasticity, on a bulk that does not reach more than 1–2 m. During their occurrence, the surface of the moving mass preserves its continuity but, at the forehead of the advancing apron, a permanent immersion of the mobilized material can be noticed, by the inversion of the herb layer. As it can also be attributed to the frost-defrost phenomenon, solifluxion was renamed as pseudo-solifluxion, because in the target area it does not possess the typical characteristics of the process (Ichim 1969, 1972, 1979).

In the Cuejdiu basin, this phenomenon is favoured by the presence of marn clay rocks specific to Hangu layers and presents a pronounced dynamic during the warm season or at the end of the cold season, when the superficial deposits are humidified. In both cases, precipitation is the mobilizing agent that determines the damping and raising of deposit plasticity. Of that, the most indicative example shows up in Sălătruc Valley (left tributary of Cuejdiu). Similar processes take place in more sectors of the Cuejdiu basin, but with preponderance in the lower sector of the main tributaries (Runc, Țiganca, Cuejdel), where the development and stabilization occur in a relatively short span of time of only 1–3 years.

Creep. Gravitational creep holds a high share in the present modeling of the versants characterized by low fragmentation and a declivity higher than 15°–20°. It is known that the same phenomenon is responsible for the cracks and ruptures in the mass of a landslide. When the modeling occurs on the slopes formed on bed heads

Table 4.1 Meteorological and pluviometric stations used for climate factors analysis

Main stations	Altitude [m]	Latitude [N]	Longitude [E]	Period of observations
Cuejdi	450	46°59′	26°17′	1970–2014
Piatra Neamț	360	46°56′	26°23′	1970–2014
Secondary stations				
Târgu Neamț	388	47°14′	26°23′	1952–2014
Pipirig	575	47°13′	26°05′	1950–2014
Pluton	660	47°12′	26°01′	1972–2014
Cracăul Negru	575	47°04′	26°13′	1973–2014
Magazia	540	47°06′	26°15′	1973–2014

in cross-valley sectors, the creep also leads to small valley formations, which take over the superficial outflow on the surface of great spread landslides.

4.3 The Role of Climate Factors

The climate database was processed according to the observations conducted between 1970 and 2014 at the meteorological station from Piatra Neamț and the pluviometric site from Cuejdi. These cumulated daily, monthly and annual values of precipitation quantities and the variation of daily, monthly and annual temperature (minimum, maximum and average temperatures). To supplement the string of data, meteorological observations made at Cuejdi by Baroi (2012), during the concurrent time interval were used.

Extending the data range towards the area from the upper basin of the Cuejdi River included the meteorological observations conducted at the meteorological stations and pluviometric sites neighboring Cuejdel Lake, specifically: Tg. Neamț, Pipirig, Pluton, Cracăul Negru and Magazia (Table 4.1).

4.3.1 The Genetic Factors of Climate

The genetic factors of climate in the studied area are generated by: the latitudinal position within Romania (the 47th parallel North), by solar radiation and by the general circulation of the atmosphere. The geographical location within the medium-high mountains from the eastern part of Eastern Carpathians influences the atmosphere through the active surface. In the Cuejdi River hydrographic basin, it enforces the altitudinal variation of climate parameters, it also influences the intensity and direction of the winds, it determines the solar rays angle of incidence through declivity and slope orientation and the degree of vegetation cover adjusts the temperature and the rainfall regime. In these conditions, according to Sandu et al. (2008), the value of solar radiation varies between 120 kcal/cm^2/year at Piatra Neamţ and 110 kcal/cm^2/year in the mountainous area corresponding to the interfluve ridges from the upper basin of Cuejdiu River.

The general circulation of atmosphere is the main cause of the acyclic variations of climate elements and corresponds with the general circulation of air above Romania, dominant being: Western circulation, Polar circulation, Tropical circulation and blocking circulation.

The Western circulation presents a 45% frequency above the country and is active in the studied area through a slight decrease in air humidity, as a result of Föhn related processes that cross the Western part of the Eastern Carpathians. Polar circulation presents a 30% frequency and is less active during winter, causing sudden cooling of weather. The tropical circulation presents a low frequency (15%), but holds great importance for the region as it determines winter warming phenomena known as "moină"—thaw, and especially for the excessive heat during summer. The blocking circulation has as a characteristic feature the deflection of atmospheric disturbances generated by the Atlantic Ocean or by the Mediterranean Sea to the North-East, causing severe episodes of drought (Sandu et al. 2008; Baroi 2012).

Atmospheric action centers that determine the mentioned circulation patterns are: the Azores Anticyclone (North Atlantic Subtropical High—NASH), the Icelandic Low, the Siberian Anticyclone, the Mediterranean Cyclone. The Azores Anticyclone is a vast area of high pressure that triggers on Romania's territory, especially in summer, the infiltration of humid air from the Atlantic which in turn determines high amounts of precipitation in that period of the year. The Icelandic Low is crucial especially in the warm season, when it allows oceanic air masses to permeate and cause cold and wet weather. In the studied area, these air masses became influenced by the Föhn, as they cross the Carpathian range and thus, rainfall quantities decrease. The Siberian Anticyclone is formed in the cold season and determines the winter frost through the immobility of air masses and sometimes even blizzards. The Mediterranean Cyclone is formed in the Southern part of Europe and through its different trajectories generates abundant quantities of precipitation in the Southern and Eastern part of the country, including in the studied area (Sandu et al. 2008; Baroi 2012).

4.3.2 Description of Main Climate Elements and Phenomena

The depiction of main climate elements and climate phenomena depends on the altitudinal zonation in which the Cuejdiu Valley is situated (minimum altitude: 306.5 m in Bistriţa Valley; maximum altitude: 1213 m Prislopaşu Peak) on the Southern aperture, on the Bistriţa corridor direction and East, towards the Subcarpathians (Cracău–Bistriţa Depression), where the penetration of continental air masses, warm in the summer and cold in the winter adjusts their regime.

Air temperature. In order to establish monthly and annual average values of air temperature data from the Piatra Neamţ meteorological station, located South-East from the studied area were analysed, as well as data from the pluviometric site Cuejdi, including on both stations 45 years of meteorological observations, starting in 1970 and until 2014. Extending the data range for the upper basin was made according to the increase of thermal gradient and was calculated on altitudinal considerations between the two stations, the value being of 0.55 °C/100 m.

The studied region was defined by average annual temperatures for the mentioned period (1970–2011), ranging between 8.9 °C at Piatra Neamţ and 8.0 °C at Cuejdi, with lower values on the highest peaks of the hydrographic basin, precisely 4–5 °C. The warmest month is July, with average temperatures of 19.6 °C at Piatra Neamţ, 18.8 °C at Cuejdi and 15–16 °C on the mountainous peaks exceeding 1000 m altitude. January is the coldest month, with average temperatures in the period mentioned of −2.1 °C at Piatra Neamţ and −3.0 °C at Cuejdi, while on the mountainous peaks it decreases to under −5 °C (Table 4.2).

The evolution of temperatures from the extreme months emphasizes the fact that in January, the lowest average temperatures for the period 1970–1979 were (−2.9 °C at Piatra Neamţ and −3.7 °C at Cuejdi), while in the decade 1990–1999 was the warmest (−1.0 °C at Piatra Neamţ and −2.0 at Cuejdi). July had the lowest average air temperature in the period 1970–1979 (19.0 °C at Piatra Neamţ and 18.1 °C at Cuejdi), while the decade 2000–2014 was the warmest (20.4 °C at Piatra Neamţ and 18.9 °C at Cuejdi).

Regarding th e annual average temperature for 10 year intervals, it can be observed that the lowest temperatures were recorded between 1970–1979 and 1980–1989 (8.6 °C at Piatra Neamţ and 7.6 °C at Cuejdi), and for the following two decades, 1990–1999 and 2000–2014, a significant increase in temperature was recorded (−2.9 °C at Piatra Neamţ and −3.7 °C at Cuejdi). The thermical variations confirm the tendency of increasing temperatures in the last 10 years, both in the mountain area and the Subcarpathian area (Baroi 2012).

The analysis of average variations in monthly minimum and maximum temperatures reveals that the thermal amplitudes recorded at both meteorological stations are higher during the winter than in the summer. The explanation is provided by the shifting of air masses during the cold season that enter the country territory without being influenced by the solar radiation which is highly reduced in winter and preserves values similar to the initial ones, while during the summer, the thermal amplitudes are smaller due to more intense solar radiation (Table 4.3).

The average level of monthly maximum temperatures recorded a difference between the station from Piatra Neamţ and the station from Cuejdi ranging from temperatures of 0.5 °C in May, 1.7 °C in December and 1.5 °C in January. The thermal variation is a result of solar radiation that presents a short period of insolation during winter, with lower values at Cuejdi station than those recorded at the meteorological station from Piatra Neamţ. For the level of monthly absolute maximum temperatures, the greatest difference between the two stations was recorded in October (3.1 °C), and the smallest in May and November. The maximum temperature recorded during the interval 1970–2014 was of 38.3 °C at Piatra Neamţ and of

Table 4.2 The average monthly temperature [°C] within the main stations of the Cuejdiu hydrographic basin between 1970 and 2014

Station	I	II	III	IV	V	VI	VII	VIII	IX	X	XI	XII	Avg
Piatra N.	−2.1	−1.0	3.0	9.1	14.4	18.0	19.6	18.9	14.3	9.2	3.5	−0.6	8.9
Cuejdi	−3.0	−1.8	2.3	8.3	13.9	17.2	18.8	18.0	13.7	8.3	2.4	−1.7	8.0
Difference	0.9	0.8	0.7	0.8	0.5	0.8	0.7	0.9	0.6	0.9	1.1	0.9	0.9

Avg Average

Table 4.3 The average maximum daily temperature [°C] within main stations of Cuejdiu hydrographic basin between 1970 and 2014

Station	I	II	III	IV	V	VI	VII	VIII	IX	X	XI	XII	Avg
Piatra N.	11.6	13.1	19.5	23.4	27.8	30.6	32.6	31.7	28.1	24.8	17.8	12.6	22.8
Cuejdi	10.1	12.1	18.2	22.3	27.2	29.7	31.5	30.4	27.1	23.6	16.9	10.9	21.7
Difference	1.5	1.0	1.4	1.1	0.5	0.9	1.0	1.3	1.1	1.3	1.0	1.7	1.1

Avg Average

37.2 °C at Cuejdi, the difference of 1.1 °C being similar as in the case of monthly maximum averages differences (Table 4.4).

The monthly minimum temperatures present higher thermal variations between the two stations during the warm season than in the cold season, the annual difference being of 2.1 °C. The phenomenon is produced by the fact that at Cuejdi the landscape configuration allows the stationary front of cold air during night, with smaller minimum values than in Piatra Neamţ. It is worth noting that the average of monthly absolute minimum temperatures in Piatra Neamţ is higher with 3.3 °C than the ones from Cuejdi, as a result of the temperature inversion phenomenon (Baroi 2012) (Table 4.5). The minimum temperature in 1970–2014 interval was recorded at Piatra Neamţ, −25.3 °C, while at Cuejdi was −26.4 °C, the difference of 1.1 °C, being preserved as in the case of absolute maximum temperatures (Table 4.6).

The monthly maximum amplitudes at both stations generally present close values, the most notable being recorded for the thermal zonation during the cold season. The maximum value of the amplitude is recorded in March (47.2 °C at Piatra Neamţ and 47.4 °C at Cuejdi), as in the first half of the month the thermal inertia mantains the winter values, while in the second part of the month, the air warms significantly and the amplitudes increase. The annual average of monthly maximum values is of 37.6 °C at Piatra Neamţ and 38.8 °C at Cuejdi, the similar values suggesting that there are not significant differences between the evolution of absolute maximum and minimum values recorded at the two stations (Table 4.7).

Air temperature for the Cuejdel Lake. The air temperature also adjusts the thermal gradient at the surface of natural barrage Cuejdel Lake. Results of annual measurements revealed that at the surface, the maximum temperature is almost 23 °C and is recorded in the summer. The same temperature is recorded on the whole surface of the lake as the river flow is insignificant and the water dynamics is reduced. During autumn, a maximum temperature of 13 °C is recorded. The high values are generated by the fact that the significant mass of water hardly releases any heat.

Frost is a meteorological phenomenon that is produced when the temperature reaches or falls bellow 0 °C. In the studied region the average date of the first frost is 14th of October at Piatra Neamţ and 1st of October at Cuejdi, while in the upper basin, in the lake area takes place in the last part of September. The average date of the last frost is 16th of April at Piatra Neamţ and 21st of April at Cuejdi, but these values can be highly exceeded at both stations. The first early frost can be recorded at second half of September while the late spring frosts can be recorded even at the beginning of May.

The minimum temperature during winter facilitates the formation of ice bridges on the surface of the lake reaching up to 25–70 cm. The depression aspect is favorable for thermal inversion that in turn produces a significant decrease in air temperatues; therefore the ice bridge can sustain for longer periods. In the winter of 2011, the maximum thickness of ice reached 37 cm, in 2012 reached 45 cm, and in 2013 reached almost 50 cm (Fig. 4.15).

Relative humidity of air. The relative humidity of air represents the degree of saturation of the air laden with water vapors and is expressed in percentages through the real tension in relation to maximum tension of water vapors to air temperature

Table 4.4 The absolute maximum temperatures [°C] within main stations of Cuejdiu hydrographic basin between 1970 and 2014

Station	I	II	III	IV	V	VI	VII	VIII	IX	X	XI	XII	Avg
Piatra N.	17.2	22.2	27.2	29.0	33.5	36.0	38.3	36.6	35.4	31.7	25.1	20.6	22.8
Cuejdi	15.4	20.0	25.0	27.1	32.8	34.6	37.2	35.6	33.6	28.6	24.4	18.8	21.7
Difference	1.8	2.2	2.2	1.9	0.7	1.4	1.1	1.0	1.8	3.1	0.7	1.8	1.1

Avg Average

Table 4.5 The average minimum daily temperature [°C] within the main stations of the Cuejdiu hydrographic basin between 1970 and 2014

Station	I	II	III	IV	V	VI	VII	VIII	IX	X	XI	XII	Avg
Piatra N.	−15.1	−13.6	−8.4	−1.1	3.8	9.1	10.5	9.5	4.4	−2.4	−7.0	−13.3	−2.0
Cuejdi	−16.3	−15.8	−10.9	−2.9	1.4	5.8	8.1	6.9	2.1	−3.9	−8.6	−14.8	−4.1
Difference	1.2	2.2	2.5	1.8	2.4	3.3	2.4	2.6	2.3	1.5	1.6	1.5	2.1

Avg Average

Table 4.6 The absolute minimum temperatures [°C] within the main stations of the Cuejdiu hydrographic basin between 1970 and 2014

Station	I	II	III	IV	V	VI	VII	VIII	IX	X	XI	XII	Avg
Piatra N.	−25.3	−21.8	−20.0	−5.3	0.3	5.4	7.6	7.2	−1.3	−6.6	−17.0	−22.0	−7.3
Cuejdi	−26.4	−23.8	−22.4	−6.0	−4.0	2.0	5.0	3.2	−5.0	−7.8	−17.6	−25.0	−10.6
Difference	1.1	2.0	2.4	0.7	4.3	3.4	2.6	4.0	3.7	1.2	0.6	3.0	3.3

Avg Average

Table 4.7 The monthly and annual amplitude of absolute minimum and maximum temperatures [°C] within the main stations of the Cuejdiu hydrographic basin between 1970 and 2014

Station	I	II	III	IV	V	VI	VII	VIII	IX	X	XI	XII	Max
Piatra N.	42.5	44.0	47.2	34.3	33.2	30.6	30.7	29.4	36.7	38.3	42.1	42.6	63.6
Cuejdi	41.8	43.8	47.4	33.1	35.8	32.6	32.2	32.4	38.6	36.4	42.0	43.8	63.9

Max Annual maximum amplitude

Fig. 4.15 Ice bridge upon the Cuejdel lake surface from winter season of 2012, 2013 and 2014

at the time the observation is made. Near aquatic surfaces and forests which are perpetual sources of evaporation and evapotranspiration, the humidity degree increases significantly (Sandu et al. 2008).

The level in the Cuejdiu Valley of relative air humidity reaches an annual average of 68% and varies between 61% (April) and 77% (November). In Piatra Neamţ, the annual average value increases with 10%, being influenced by the Eastern aperture of the Bistriţa corridor towards the Subcarpathian area. The significant differences between the upper section of the Cuejdiu Valley and the Bistriţa Valley are generated by the Eastern influence of air masses, the decrease of vertical gradient of relative humidity that varies between 0.5 and 0.9%/100 m, the increased degree of cover and by the greater frequency of Föhn processes at higher altitudes (Apăvăloae et al. 2001; Apostol 2004) (Table 4.8).

Nebulosity and sunshine duration. The atmospheric nebulosity is in direct dependency with particularities of general circulation of air masses, with thermal stratification of the amosphere and with the convective movements of the atmosphere but also depends on the landscape configuration. Taking into consideration the morphographic and morphometric conditions of the area, the majority of cloud structures are formed as a result of the orographic convection. The total nebulosity increases due to altitudinal considerations with approximately 0.3–0.4 tenths/1000 m, and reaches

Table 4.8 The average monthly humidity of air [%] within the main stations of the Cuejdiu hydrographic basin between 1970 and 2014

Station	I	II	III	IV	V	VI	VII	VIII	IX	X	XI	XII	Avg
Piatra N.	80	79	76	73	74	75	76	77	79	80	82	82	78
Cuejdi	72	71	68	61	63	62	62	62	67	72	77	76	68

Avg Average

in the Subcarpathian sector 5.8–6.2 tenths, and in the higher area (>1000 m), values between 6.3 and 6.6 tenths (Apostol 2004) (Table 4.9).

In such conditions, the sunshine duration records moderate values in comparison with rest of the country and even smaller within the Moldavian Subcarpathians area due to the increase of nebulosity in the Bistrița Valley. In Piatra Neamț, the annual maximum value for the analised interval is 2293.5 h, and the minimum is 1705.9 h and were recorded in 1996. The multiannual average is situated around the value of 2018.4 h (Apăvăloae and Apostol 2002) (Table 4.10).

In the Cuejdel Lake area, situated at the relative altitude of 661 m, the total nebulosity values and the sunshine duration present few differences compared to the meteorological stations located downstream in the Cuejdiu Valley. The differences are generated by the slope position and the mountainous peaks dominating the lacustrine unit, where the topoclimate generated by the slopes facing North and North-East are extended more and the slopes facing West and South-West are shorter, but have a greater degree of declivity. During clear days, mornings are shadier on the left bank, the sun rays reaching the water level with approximately 30 min later than the right bank and during the afternoon the right bank is shady and the left bank is sunnier for approximately 1 h.

Atmospheric precipitations. Precipitations represent, alongside temperature, an essential climate factor in the evolution and limno-ecological condition of Cuejdel Lake. Their analysis was made based on data from the Piatra Neamţ meteorological station and the pluviometric site from Cuejdi, collected in 45 years of research (1970–2014). Although the two stations describe only the precipitation regime from the lower and medium part of the studied hydrographic basin, an interpolation of data was also attempted for the upper basin based on the altitudinal gradient of increasing precipitation amount.

The annual precipitation amount recorded differs greatly at the two stations, in terms of multiannual average values and also of monthly average values. The multiannual average of precipitation quantities is 621.0 mm in Piatra Neamţ and 738.7 mm in Cuejdi, with an average amplitude of 117.7 mm generated by the local landscape conditions that amplifies the precipitation potential of the air masses on the Eastern structure. This is the explanation for the more significant quantities of atmospheric precipitations recorded at the pluviometric station in Cuejdi and in the upper hydrographic basin (Table 4.11).

In order to extend the string of quantitative data recorded in the Cuejdel Lake area, situated at an altitudinal difference of 200 m from the pluviometric station, multiannual average quantities from the following pluviometric stations were taken into consideration: Tg. Neamţ, Pipirig, Pluton, Cracăul Negru, Magazia, Cuejdiu and Piatra Neamţ, between 1970 and 2014.

The lowest value of annual precipitations was recorded at both stations in 1986 and was 443.0 mm at Cuejdi and only 362.9 mm at Piatra Neamţ, while the maximum annual value was recorded in 2010, 911.2 mm at Piatra Neamţ and 1015.2 mm at Cuejdi. For 10 years intervals, it can be observed that at the both stations, the most humid period was between 1971 and 1980, with an average of precipitation quantities between 672.5 mm at Piatra Neamţ and 804.9 mm at Cuejdi, while the driest was

Table 4.9 The average monthly nebulosity [tenths] within the area of the Cuejdiu hydrographic basin between 1970 and 2014

Station	I	II	III	IV	V	VI	VII	VIII	IX	X	XI	XII	Avg
Cuejdi	6.1	6.4	6.2	6.0	6.0	5.8	5.1	4.6	4.7	5.5	6.3	6.6	5.8

Avg Average

Table 4.10 Mean monthly sunshine [hours] within the area of the Cuejdiu hydrographic basin between 1970 and 2014

Station	I	II	III	IV	V	VI	VII	VIII	IX	X	XI	XII	Avg
Piatra N.	90	103	129	163	229	236	264	251	193	150	90	72	1969

Avg Average

Table 4.11 The monthly average quantities of precipitation [mm] recorded at the main stations of the Cuejdiu hydrographic basin between 1970 and 2014

Station	I	II	III	IV	V	VI	VII	VIII	IX	X	XI	XII	Avg
Piatra N.	20.6	20.0	29.4	53.4	78.0	96.7	103.8	78.4	49.5	39.4	27.9	24.0	621.0
Cuejdi	25.1	25.4	37.1	68.6	93.3	116.5	115.9	95.3	59.4	43.4	31.9	26.7	738.7
Difference	4.5	5.4	7.7	15.2	15.3	19.8	12.1	16.9	9.9	4.0	4.0	2.7	117.7

Avg Average

between 1981 and 1990, when the average at Cuejdi was 668.1 mm and at Piatra Neamţ 549.4 mm.

It is interesting to note that between 1970 and 1980, June was the rainiest period of the year, but in the following decades (1980–2014), July became the month with the highest precipitation amounts. The situation is slightly abnormal, since the climatological studies indicate June as being the rainiest time for this region of the country (Sandu et al. 2008). Smaller precipitation amounts are recorded at both stations during winter, the driest being February, with total average values between 20.0 mm at Piatra Neamţ and 25.1 mm at Cuejdi (Baroi 2012).

The differences between the monthly average quantities recorded at Piatra Neamţ and Cuejdi are smaller in the cold season and higher in the warm season due to the convective aspect of the precipitations that manifest locally, sometimes generating considerable differences between the two stations.

Within the seasonal distribution of rainfall, it can be observed that almost half of annual quantity falls during the summer, as a result of convective phenomena (Piatra Neamţ—2789 mm or 44.9% from the total; Cuejdi—330.9 mm or 45% from the total), while the lowest quatity is recorded during winter (Piatra Neamţ—64.6 mm or 10,4% from the total; Cuejdi—78.0 mm or 10.6% from the total). Divided into semesters, it can be observed that 74% from the annual precipitations fall in the warm season, more exactly between May-September (459.8 mm at Piatra Neamţ; 544.4 mm at Cuejdi), and the rest during the cold season, when at both stations similar values are recorded (161.3 mm at Piatra Neamţ; 191.6 mm at Cuejdi).

The analysis of monthly minimum precipitation amounts underlines that during the studied period (1970–2014) there was no month without precipitations, the lowest difference between the two stations being recorded in January (0.2 mm), and the highest in June (16.3 mm). The ratio between the monthly average quantity and the minimum values in January reveal these are 10–12 times smaller, while during summer months, the ratio ranges between 2 and 3.5 (Table 4.12).

The lowest quantity of precipitation was recorded concurrently at both stations in November 2011, respectively 0.9 mm at Piatra Neamţ and 1.4 mm at Cuejdi. It was also observed that throughout the year, the monthly minimum quantities are more abundant at the pluviometric site from Cuejdi than those recorded at Piatra Neamţ. The frequencies of months when the lowest quantities fall define January as the month with the lowest quantity recorded at both stations, respectively 32.5% at Piatra Neamţ and 37.5% at Cuejdi, while the increased frequency of monthly minimum values is recorded in January, with values between 17.5% at Piatra Neamţ and 25% at Cuejdi (Table 4.13).

The monthly maximum quantities of precipitation reveal that in the cold season, with the notable exception of March, the quantities do not exceed 100.0 mm, while during the warm season this value exceeds the threshold of 200.0 mm and sometimes 300.0 mm. An exceptional situation happened in 1991, when in July there were 317.4 mm at Piatra Neamţ and 346.3 mm at Cuejdi (Table 4.14).

The frequency of months in which the monthly maximum quantities are recorded points to a series of differences between the two analised stations. The month of July has the highest frequency, dominating the maximum pluviometric values for 15 years

Table 4.12 The monthly minimum average quantities of precipitation [mm] recorded at the main stations of the Cuejdiu hydrographic basin between 1970 and 2014

Station	I	II	III	IV	V	VI	VII	VIII	IX	X	XI	XII	Total
Piatra N.	1.8	2.4	3.6	7.1	9.4	31.2	31.7	15.5	3.9	2.7	0.9	1.3	362.9
Cuejdi	2.0	7.7	6.7	11.4	14.5	47.5	34.1	23.1	6.7	6.9	1.4	5.4	443.0
Difference	0.2	5.3	3.1	4.3	5.1	16.3	2.4	7.6	2.8	4.2	0.5	4.1	80.1

Table 4.13 The frequencies of months with the lowest quantities of precipitation [mm] recorded at the main stations of the Cuejdiu hydrographic basin between 1970 and 2014

	Station	I	I	II	III	IV	V	VI	VII	VIII	IX	X	XI	XII
No. months	Piatra N.	14	7	4	–	–	–	–	–	1	5	3	7	14
	Cuejdi	16	5	–	–	–	–	–	–	3	2	5	10	16
Total [%]	Piatra N.	32.5	17.5	10.0	–	–	–	–	–	2.5	12.5	7.5	17.5	32.5
	Cuejdi	37.5	12.5	–	–	–	–	–	–	7.5	5.0	12.5	25	37.5

Table 4.14 The monthly maximum average quantities of precipitation [mm] recorded at the main stations of the Cuejdiu hydrographic basin between 1970 and 2014

Station	I	II	III	IV	V	VI	VII	VIII	IX	X	XI	XII	Total
Piatra N.	48.6	58.4	131	133	208	260	317	218	124	117	55.8	57.4	911.2
Cuejdi	55.6	70.0	161	194	255	296	346	228	174	125	84.6	55.3	1074
Difference	7.0	11.6	29.1	61.2	46.8	35.6	17.4	8.8	50.3	8.6	28.8	2.1	162.6

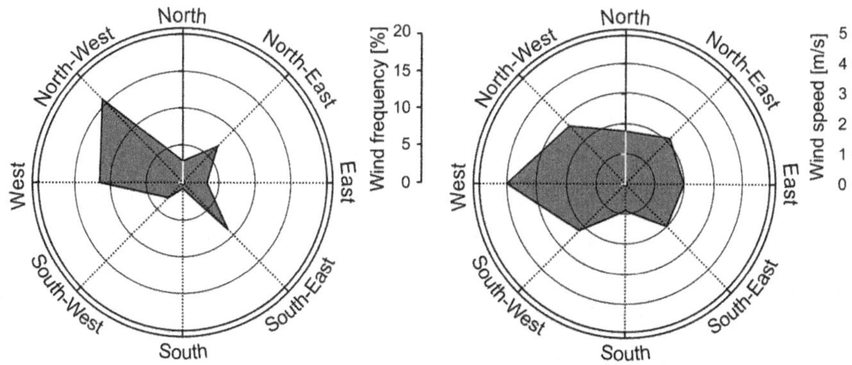

Fig. 4.16 The annual variability of wind frequency [%] and speed [m/s] recorded at the main stations of the Cuejdiu hydrographic basin (1970–2014)

at Cuejdiu and for 12 years at Piatra Neamţ. The summer months (June, July and August) have the largest share, even if considerable quantities can fall also in autumn due to the atmospheric fronts that are specific for the warm season. In conclusion, the precipitation amounts in the hydrographic basin of Cuejdi River certifies that the precipitation regime in the area is in excess (Table 4.15).

Evapotranspiration. Evapotranspiration represents a very important climate parameter for evaluating the regime of surface waters. The main factors in determining the evaporation quotient variation at the active surface level are meteo-hydric, physiological and edaphic. For a hydrographic basin, evapotranspiration is produced on all these levels, both on the topographic surface covered or not by vegetation and on the hydrographic network or bodies of water. Regarding lakes, the value of potential evaporation plays a major role in influencing the hydrological balance. For the Eastern Carpathians, the average value of evaporation for lakes is between 600 and 620 mm/year (Gâştescu 1963) (Table 4.16).

For the Cuejdi hydrographic basin, calculating the potential evapotranspiration was realized along with the pedological studies made on the territory of Gârcina Commune, gathering observations from the period 1975–1998. The data analysis shows that in the cold semester of the year (December–February), the evapotranspiration values are insignificant and in the vegetation season, they reach higher values for summer months (June, July, August); nevertheless, the pluviometric regime in surplus ensures an average water reserve that exceeds 100 mm/year.

Wind. The air masses circulation from Cuejdiu Valley depend on the general circulation of air masses specific for Eastern part of Romania. At the Cuejdi site the most frequent winds are the North-West winds (15.9%), West (12%) and South-East (8.4%), and the highest is specific for the North-West winds (2.6 m/s), North-East winds (2.2 m/s) and South-East (2.2 m/s). From the studied interval, over 45% of time there is atmospheric calm due to shelter conditions. At altitudes higher than 900–1000 m, the wind speed increases considerably, the dominant direction being West and North-West (Fig. 4.16).

Table 4.15 The frequencies of months with the highest quantities of precipitation [mm] recorded at the main stations of the Cuejdiu hydrographic basin between 1970 and 2014

	Station	I	II	III	IV	V	VI	VII	VIII	IX	X	XI	XII
No. months	Piatra N.	–	–	–	7	11	12	8	2	1	–	–	–
	Cuejdi	–	–	2	5	9	15	6	3	1	–	–	–
Total [%]	Piatra N.	–	–	–	17.5	27.5	27.5	20.0	5.0	2.5	–	–	–
	Cuejdi	–	–	5.0	12.5	22.5	35.0	15.0	7.5	2.5	–	–	–

Table 4.16 The monthly potential evapotranspiration [mm/year] and water reserve [mm/year] calculated using quantities of precipitation [mm] recorded at the main stations of the Cuejdiu hydrographic basin between 1970 and 2014

Cuejdi station	I	II	III	IV	V	VI	VII	VIII	IX	X	XI	XII	Avg
P Evap.	–	–	10	47	88	114	129	113	75	38	10	–	624
W Res.	22.6	24	17.3	2.1	–	0.9	–	–	–	1.9	26.1	27.1	125

P Evap. Potential evapotranspiration, *W Res*. water reserve, *Avg* Average

Fig. 4.17 Cuejdi hydrometric station (September 2016)

Alongside the general dynamic of air masses there are also local winds spawned by thermal or mechanic disturbances generated by active surface particularities. Some have a reduced frequency (Föhn, Crivăț, Suhovei) and others are dominating only a certain area (the mountain-valley winds). Usually, these winds enter in Cuejdiu Valley through the Bistrița corridor and are felt when the atmosphere circulation is slow and in some cases is overwhelmed by these winds (Sandu et al. 2008; Baroi 2012).

4.4 The Role of Hydrological Factors

The hydrological database was purchased with the support of Administration of Basin Waters Siret (ABA)—Water Management System Neamț (SGAN). The data included observation of liquid flow and daily, monthly, annual average levels recorded at the hydrometric station located on the Cuejdiu River, at approximately 600 m upstream from the confluence with Slatina Creek (46°59′36″ lat. N; 26°17′42″ long. E). For liquid flows, the period studied was 1978–2011, and the level measurements were made between 1992 and 2001 (Fig. 4.17; Table 4.17).

Unfortunately, for the hydrographic network from the upper basin of the river and for the Cuejdel lacustrine basin, there was no information regarding the hydrometric activity. All data in the present study have been interpolated based on parameters that influence liquid drain and expressed as probable values for exceeding the flows for different periods of time (for example: $Q_1\% = 100$ years; $Q_2\% = 50$ years; $Q_5\% = 20$ years).

Table 4.17 The main characteristics of the Cuejdi hydrometric station

Regional hidro-graphic basin	Hydrometric station	River	Length [km]	Hidrographic basin surface [km^2]	Locatin of hydromet-ric station	Period of observa-tions
Bistriţa	Cuejdi	Cuejdiu	15.2	57	Left bank—ter-race of 9–12 m	1978–2011

4.4.1 Underground Waters

The groundwater presents a rather continuous development in the Subcarpathian area and a more fragmented one in the moutainous area where the distribution of underground waters depends on the thickness of superficial deposits or on the soil features. Their quality is especially good and the depth of their location exceeds 5 m for most of the time.

The distribution on depth classes of groundwater level on the Cuejdiu Valley shows that in proportion of >80% it is located between 5 and 15 m, and the rest of 20% corresponds to depths of 3–5 m (cca. 10%), 2–3 m (cca. 5%), 1–2 m (cca. 1%) and 0–0.5 (cca. 0.5%). The shallow depth of the groundwater generally corresponds to the first terraces of Cuejdiu River, increasing towards the upper part of the ridges or towards the mountainous area. At higher altitudes, the temporary aquifer layers can be traced only in the profiles of soils that have been affected by pseudo-gleyzation, their presence being conditioned by the pluviometric excess specific to transitional seasons and by the presence of impervious lithological deposits at the base (Studiu Pedologic–Teritoriul Comunei Gârcina 1998).

Coast springs are frequently documented throughout the studied area, with the highest frequency at the base of mountain peaks or on the dipair contact between the Miocene lithological deposits. The sodium chloride springs from Slatina Creek and from the Eastern rim of Cozla Hill can be mentioned as suggestive examples.

4.4.2 Surface Waters

The most important role in forming the water resources is played by the landscape, both directly, through the degree of fragmentation and presence of negative forms generated by the slope and exposition or the lithology of substrate and indirectly, through climate elements zonation and implicitly the value of surface drainage and infiltration.

Regarding the formation of the studied lacustrine depression, the cumulation of all these factors led to slope processes that filled the Cuejdel Valley with deluvial material. In turn, the creek provided the necessary volume of water accumulated

Table 4.18 Summary of drainage basin parameters for the Cuejdi river basin

Drainage basin parameter	Value	Unit of masurement
Surface	97.41	km^2
Surface of left bank	61.96	km^2
Surface of right bank	35.53	km^2
Lowest altitude	306	m
Mean altitude	704.5	m
Highest altitude	1213	m
Spring river altitude	731.2	m
Perimeter length	64.75	km
Mean altitude of perimeter	948.76	m
Mean slope of perimeter	29.96	%
Sinuosity coefficient of perimeter	2.27	–
Axial length	22.71	km
Mean length	16.02	km
Axial width	7.96	km
Mean width	6.08	km
Mean width of left bank	3.86	km
Mean width of right bank	2.21	km
Main river length	29.11	km
Mean slope of main river	14.11	%
Asymmetry coefficient	0.54	–
Form factor	0.18	–
Circularity ratio	0.28	–
Elongation ratio	0.48	–
Compaction coefficient	1.85	–

behind the dam, consequently controlling the input and output from the lake through drainage. This is another reason why the particularities of the hydrological basin will be analysed at the level of Cuejdel hydrographic basin in which the drainage basin of Cuejdel Lake is integrated.

The hydrographic network. In general, the hydrometric features from the studied area are specific to medium size rivers formed in the North-Eastern Carpathian area and tributaries to Bistriţa River. The Cuejdi River basin consists of a well developed drainage area in the upper part of the basin from where it collects its main tributaries and becomes straighter in the lower sector where the valley narrows, concentrating the drainage on the main riverbed (V Order—according to Horton-Strahler classification). This way of arrangement of the hydrographic network within the basin is framed under the dendritic type (Table 4.18).

Table 4.19 Relation between the network order and the number of river segments within the Cuejdiu hydrographic basin

Drainage network order (Horton-Strahler)	Number of river segments	Relative frequency of river segments [%]	Length [km]
I	422	49.01	142.66
II	220	25.55	72.83
III	73	8.47	19.52
IV	50	5.8	14.22
V	96	11.1	29.71
Total	861	100	278.94

Table 4.20 The slope of main river (I and II Order) within the Cuejdiu hydrographic basin

Drainage network order (Horton-Strahler)	River	Length [km]	Lowest altitude [m]	Highest altitude [m]	Slope [%]
IV	Pascău	1.32	666.94	767.98	76.54
IV	Ţiganca	2.72	530.93	649.01	43.41
IV	Cerbu	2.22	717.11	799.96	37.31
IV	Cuejdel	3.79	575.92	702.34	33.59
IV	Gârceniţa	2.43	398.34	461.65	26.05
IV	Runcu	2.99	488.16	553.92	21.99
V	Cuejdiu	29.11	306.5	717.35	14.11

The I and II Order elementary valleys contain approximately 77% of the total length of hydrographic network, and III and IV Order make up for 12%, and the main stream (V Order) the rest of 11%. The dominance of I Order type of hydrographic network is also obvious in the number of river sections that accounts for 50% of the total number. It is worth noting the higher number of river section belonging to V Order (96) compared to those belonging to IV Order (73) and IV Order (50). The explanation is provided by the elongated shape of the hydrographic basin that generates narrow slopes that, in turn, manage to form only a torrential drainage network (I and II Order) on some sections. (Table 4.19).

The IV and III Order basins develop within the landscape where the main valley is wider, respectively in the upper section on the right slope (Runcu Peak) and on the left slope (Cerbul Peak, Cuejdel Peak, Ţiganca Peak). The sector where the valley presents asymmetric slopes generates superior types of valleys (III and IV), on the bank with reduced declivity due to the favourability of deep erosion processes (Gârceniţa Peak, Boroda Peak) (Fig. 4.18; Table 4.20).

The main stream has an average slope of the riverbed of 14.11%, and the arrangement in longitudinal profile indicates an evolutionary stage that easily surpassed the morphological balance phase. In contrast, the Cuejdel Creek, its main tributary, holds

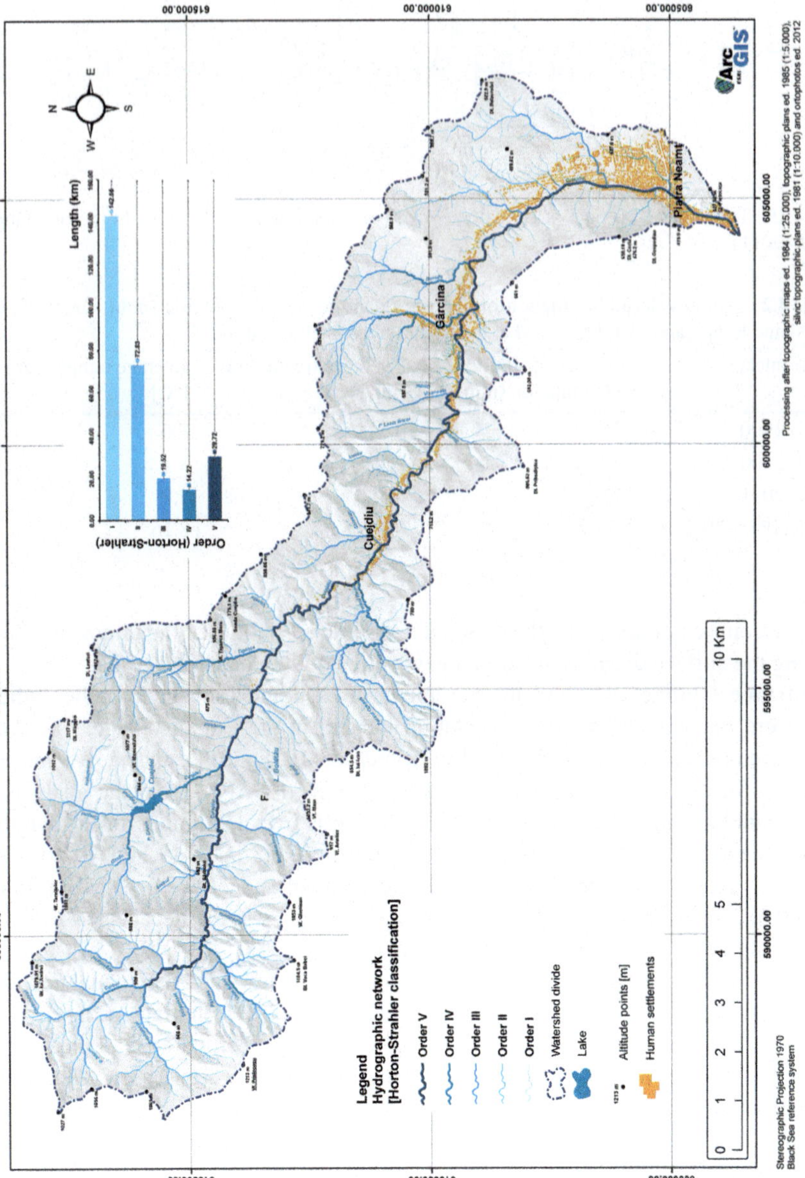

Fig. 4.18 Map of the hierarchy of hydrographic network according to the Horton-Strahler system within the Cuejdi river basin

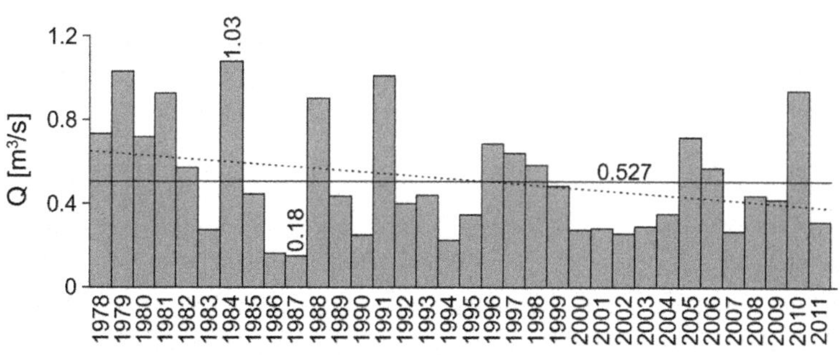

Fig. 4.19 The annual average stream flow [m^3/s] recorded at the Cuejdi hydrometric station (1970–2011)

Table 4.21 Annual decades variation of average quantities of precipitation [mm], stream flows [m^3/s] and temperature [°C] recorded at the Cuejdi hydrometric station

Time interval	Average quantities of precipitation [mm]	Average stream flow [m^3/s]	Average temperature [°C]
1981–1990	668.1	0.518	7.6
1991–2000	702.1	0.504	8.1
2001–2011	750.4	0.470	8.8
Total (1978–2011)	706.8	0.497	8.16

a convex curve influenced by the deluvial mass that reconfigured the minor riverbed, forcing the surface drainage to form a new valley on the landslide surface.

Surface drainage. The only measurements of the liquid drainage from the studied area were made at the hydromeric station located in the Cuejdi village. The data range used by us is placed within a 34 year interval (1978–2011), and were provided by ABA Siret—Water Management System Neamț (SGAN). From the analysis a multiannual average stream flow of 0.527 m^3/s was revealed, much lower than the values mentioned in the professional literature (app. 0.8 m^3/s) (Ichim 1979). The averages on ten yearly show a continuous decrease of annual average stream flows from 0.607 m^3/s between 1978 and 1987, to 0.529 m^3/s in the decade 1977–1988, decreasing to 0.465 m^3/s in the last 14 years (1998–2011) (Fig. 4.19).

The correlation between the annual average stream flows and the annual precipitations reveals the fact that there is no straight relation between them, as a series of dirsturbances intervene, such as the torrential feature of precipitation and the evapotranspiration determined by the high temperatures from the warm season. Therefore, the annual average drainage regime recorded in the Cuejdi River area is in perpetual decrease due to sudden increases of average temperatures (7.6 °C = 1978–1987; 8.1 °C = 1988–1977; 8.8 °C = 1998–2011), while the sudden and short increases of flows are generated by the torrential downpours (Table 4.21).

In longitudinal profile, the biggest declivities are specific for the most of secondary valleys because it drains the upper part of the slopes. Along these slopes, the trickling happens in short amount of time, generating torrents that overfill in the superior order

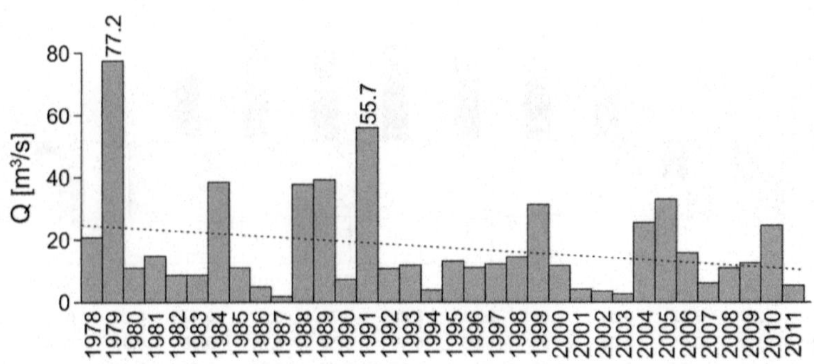

Fig. 4.20 The annual maximum stream flow [m^3/s] recorded at the Cuejdi hydrometric station (1970–2011)

valleys (III, IV and even V) during periods with high pluviometric values. At a closer examination made on the IV order arteries it was observed that the highest declivity belonged to Pascău Creek (76.54%), formed on the Northern slope of Prislopaş Peak. In a similar manner, the longitudinal representation of Ţiganca Creek (43.41%), Cerbu (37.31%) and Cuejdel (33.59%) are formed, indicating considerable average declivities and confirming an accelerated speed of liquid trickling.

The average annual streamflows exceeded the value of 1.0 m^3/s in just three years from the studied period: 1.03 m^3/s in 1979, 1.08 m^3/s in 1984 and 1.0 m^3/s in 1991, when precipitation amounts exceeded the multiannual average value (917.1 mm in 1979; 880.8 mm in 1984; 1052.3 mm in 1991). Year 2010 stands out as the precipitation quantities reached 1072.5 mm, the annual average flow was only 0.936 m^3/s. The annual maximum stream flow chart reveals once more that the drainage is in a straight correlation with rainfall only when these take place in a short period of time. Year 1979 also stands out as then the maximum maximorum flow was recorded—77.2 m^3/s (05.VIII.) and also 1991 with a flow of 55.7 m^3/s (04.VII.) (Fig. 4.20).

The monthly drainage regime bears the imprint of precipitations regime either with torrential feature or solid or liquid form in different times of the year. In addition, other factors induced by the evapotranspiration, by the underground water reserves or by the degree of vegetation cover play a significant role.

The monthly average flows correspond to a variation zonation comprised between the minimum of 0.20 m^3/s in January and the maximum of 1.09 m^3/s in April when the ice melts. It can be observed that the extreme flows recorded during the summer months have an irregular frequency and as a result the highest monthly average flows are formed by the spring liquid drainage. Conversely, the monthly maximum average flows reveals the torrential feature of the summer months, with a distribution of the highest values of this parameter being recorded in June (6.46 m^3/s), July (5.56 m^3/s) and August (6.71 m^3/s) (Fig. 4.21).

The minimum average flows hold the lowest values during winter months, when liquid precipitations are greatly reduced in quantity and the drainage is reduced almost to the freezing point (December, January, February—0.11 m^3/s). In the warm

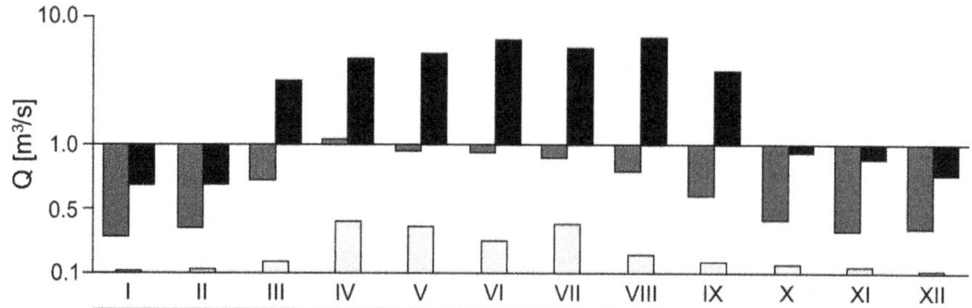

	I	II	III	IV	V	VI	VII	VIII	IX	X	XI	XII
Avg.	0.20	0.22	0.53	1.09	0.91	0.87	0.79	0.63	0.40	0.26	0.22	0.22
Max.	0.48	0.50	3.06	4.64	5.05	6.46	5.56	6.71	3.68	0.87	0.78	0.58
Min.	0.11	0.11	0.16	0.38	0.35	0.27	0.37	0.20	0.16	0.14	0.13	0.11

Fig. 4.21 The monthly average, maximum and minimum stream flow [m^3/s] recorded at the Cuejdi hydrometric station (1970–2011)

season, they do not exceed the limit of 0.40 m^3/s and correspond to the drought-afflicted periods.

The seasonal distribution of average drainage demonstrates that winter holds the lowest place in the annual drainage (10.1%), preceded by the autumn drainage (13.9%). Spring amounts to 40.3% and summer 35.7%, effectively dominating the annual drainage. Distributed on semesters, during the first half of the year, January-June, drainage represents 60% of the total value and the II semester the rest of 40%.

The most important parameter in analysing the hydrological factors that can modify the morphology of lacustrine depression and the limno-ecological parameters of the Cuejdel Lake is represented by the maximum drainage associated with floods and overflows. Their genesis is directly linked to the existent climate conditions in the studied hydrographic basin and can be determined by the superficial drainage caused by rains (pluvial floods), by the sudden melting of snow (nival floods), by the superposition of the two phenomena (pluvial-nival floods) or as a result of an anthropic intervention in the hydrographic network or lakes (Zăvoioanu 1978; Minea and Romanescu 2007; Romanescu 2000, 2009a, b, c).

In the hydrographic basin of Cuejdiu River, the most negative phenomena associated with maximum drainage between 1978 and 2011 were the huge quantities of rainfall recorded in the warm season in the following years: 1978, 1979, 1984, 1988, 1989, 1991, 1999, 2004, 2005 and 2010 (Table 4.22).

The maximum maximorum flow recorded at the hydrometric site Cuejdi was 77.2 m^3/s and took place on 15th of August 1979, one year from the first gravitational displacements that blocked the riverbed of Cuejdel Creek. It is supposed that this flood was made up of two successive peaks, first between 21st and 25th of June (Q max. 45.2 m^3/s) and the second between 13th and 20th of August (Q max. 77.2 m^3/s), that carried the deluvial material accumulated in the Cuejdel riverbed, releasing the water volume behind the barrage and effectively destroying the first lake.

Table 4.22 Major floods occurred at the Cuejdi hydrographic station between 1978 and 2011

Hydrometric station	River	Date of occurrence of flood peaks	Q [m^3/s]	Percentage of the maximum flow [%]
Cuejdi	Cuejdiu	30.VI.1978	20.8	26.9
		15.VIII.1979	77.2	100
		25.IV.1984	38.1	49.3
		29.III.1988	37.5	48.5
		06.IX.1989	39.0	50.5
		04.VII.1991	55.7	72.1
		28.VII.1999	31.2	40.4
		05.VIII.2004	25.4	32.9
		12.VII.2005	32.9	42.6
		26.VI.2010	24.6	31.8

The second notable event took place in 1991, when historical flows were recorded on most of Romania's rivers due to excessive rains. Between 1st and 10th of July, on the Cuejdel River a first overflow with a maximum flow of 18.6 m^3/s took place. After a relatively calm period, from a hydrometric point of view, between 28th of July and 5th of August a new overflow with two peaks, first reaching on 29th of August the value of 55.7 m^3/s and the next one, on the 3rd of August, with a value of 24.8 m^3/s took place. The negative hydrographical events associated with maximum drainage from that year corresponds to the formation of the Cuejdel Lake (Fig. 4.22).

In addition to these events, similar floods produced by the Cuejdi River were recorded, on the 7th of June 1975, when the estimated flow was 100 m^3/s at Cuejdi and over 150 m^3/s at Piatra Neamţ where, due to a bridge being blocked, water overspilled, flooding most part of Dărmăneşti neighbourhood and the surrounding houses (Baroi 2012).

4.5 The Role of Bio-pedogeographical Cover

The bio-pedogeographical database includes information from the pedological studies made for the unforested territory of Gârcina commune and Piatra-Neamț municipality. Depicting the pedoedaphic cover within the forest space belonging to the research area was made based on forestry studies between 1950 and 2005 and published in "Forestry improvement of Gârcina Forestry District". The evolution of forest surfaces was documented from the same sources.

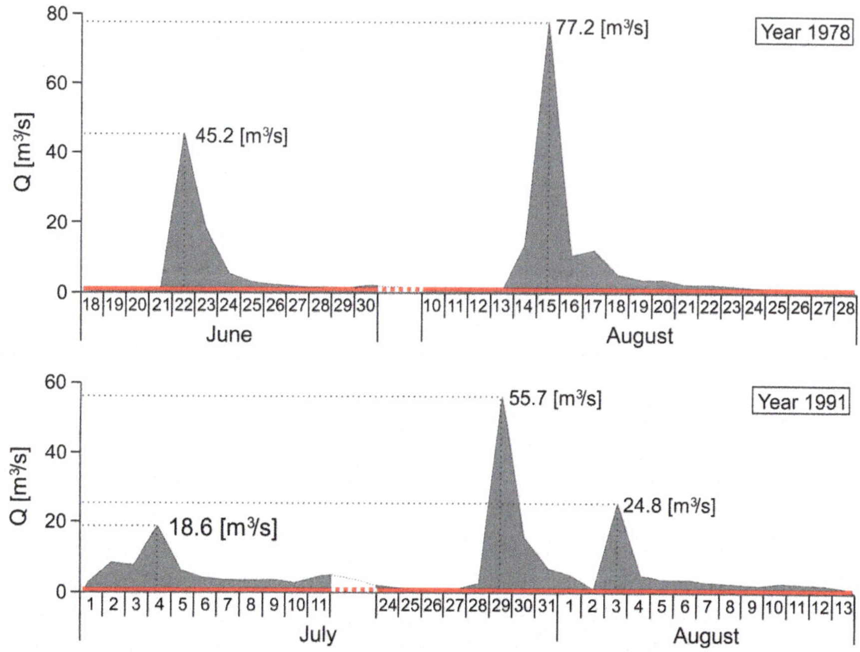

Fig. 4.22 Hydrograph of major floods occurred at the Cuejdi hydrographic station in years 1978 and 1991; the red line represents the multiannual base flow for June, July and August

4.5.1 Land Use

The particularities of bio-pedogeographical cover from the Cuejdi hydrographic basin hold an important significance in preserving the lacustrine unit Cuejdel through its high degree of slope forestation (73.29%), which in turn adjusts the drainage regime on the slope, as well as the soil features that control the infiltration of water from rainfall.

Consulting the Forestry Improvement of the Gârcina commune made in 1950, 1963, 1974, 1985, 1995 and 2005, one can observe the fact that the forest dominates over 70% of the studied area surface, and the forest stock is predominantly made up of three main tree species: beech (35–41%), fir (22–36%) and spruce (15–20%). The rest (app. 9–22%) consists of other wooden species, most frequent being cluster of oaks, hornbeam and sycamore. The same source shows that the forests on the Cuejdi Valley are divided in two production units, one located on the right slope, called UP Runcu and the other on the left slope, managing the forested areas in the lake area called UP Mijlocul (Amenajamentele Silvice ale Comunei Gârcina 2005).

The distribution of tree species on the two UPs indicates the predominance of beech tree in both production units, while the coniferous dominates the right slope of the valley through a higher rate of fir and spruce. The oak is present in the UP Mijlocul, with a high percentage (15%), especially in the middle and upper part of

Table 4.23 The relative frequency [%] of tree species in forest Production Units [UP] of the Cuejdiu Valley in 2005

Production Unit	Beech	Fir	Spruce	Oak	Sycamore	Other hardwood	Other softwood
Runcu [%]	35	36	20	1	3	3	2
Mijlocul [%]	41	22	15	12	3	6	1

Fig. 4.23 Forest types around Cuejdel lake—views from different points of the catchments basin

the sunny slopes to the South, while at the lower part, close to 500 m altitude, the spruce descends. The species that are rarely represented appear in a mixture of the already mentioned deciduous and coniferous species, especially the sycamore, with a higher frequency on the left bank but also the hornbeam covering the Subcarpathian hills Mintiana and Balaurul (Table 4.23; Figs. 4.23 and 4.24).

The distribution of forests on altitudinal classes emphasizes the interval 600–800 m, 39% of total surface, followed by the 800–1000 m interval, 28% and the 400–600 m, 29%. At over 1000 m altitude, the forest amounts to 5% from the total surface, and the rest 2% is made up of surfaces situated under 400 m, where land clearings have been made to build settlements. Within this distribution, the altitudinal zonation avoids the accustomed vertical disposing as a result of local climate particularities favoring the oak settlement at 800–1000 m altitude and the descent of the fir tree close to 500 m altitude (Fig. 4.24).

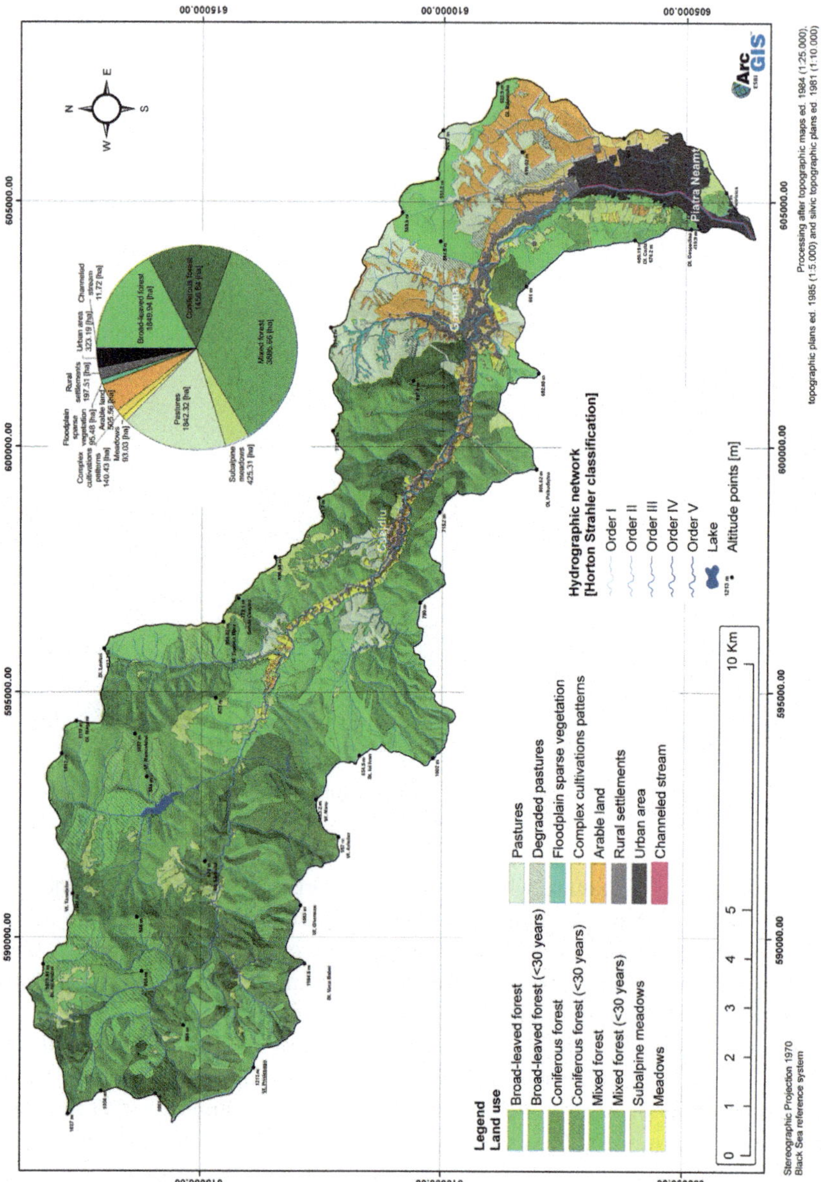

Fig. 4.24 Land use map of Cuejdiu hydrographic basin

Table 4.24 Evolution of forest areas [ha] within the Cuejdiu valley between 1963 and 2005

Production Unit	Year					Multiannual average
	1963	1974	1985	1995	2005	
Runcu [ha]	3095	3097	3097	3117	3073	3095.8
Mijlocul [ha]	3437	3647	3662	3662	3595	3600.6

Table 4.25 Timber exploitation amount [m^3] in the Cuejdiu valley between 1963 and 2005

Production Unit	Year					Multiannual average
	1963	1974	1985	1995	2005	
Runcu [m^3]	475,310	232,830	198,213	200,165	152,151	25,173.88
Mijlocul [m^3]	358,400	320,990	298,550	162,474	133,470	254,766.8

In the area of Cuejdel Lake, the forest stock experienced weak dynamics concerning the forested area, with a slight improvement between 1963 and 2005 with 158 hectares in the Cuejdel Valley also seen from the amount of exploited wood, from 358,400 m^3 between 1950 and 1963, to 92,651 m^3 in 2005–2010 (Tables 4.24 and 4.25). Another aspect worth of attention is the age of forests surrounding the lake, where between 1995 and 2004 a series of afforestations and planting campaigns were made amounting to 61.4 hectares, out of which 40.7 ha (66.3%) were integral afforestations and 20.7 ha (33.7%) completions. On the left slope a reforested area with fir and spruce appeared bordering the landslide under the Muncel ridge to the North (Fig. 4.24).

The forest vegetation is directly responsible for all soil types in the region and their distribution is identical for the most part with the land use. Based on ortophotoplans made after 2005 flights, the main elements specific for the distribution of main uses of land within the Cuejdi hydrographic basin were identified. There were 13 types of land use identified, among which, as it was expected, forests hold the main proportion, amounting to 73.29% of the total surface, out of which: 3886.66 ha mixture forest, 1849.94 ha decidous forest and 1456.84 ha coniferous forest (Fig. 4.25).

With only 26.71% from the total surface, the rest of land use types are grouped in the lower basin along the main valley or on the Subcarpathian milder slopes and with a favorable substrate for agriculture. Among these, the mountain meadows (559.78 ha) and lands used as pastures surrounding human settlements (582.2 ha) are emphasized. An important percentage is held by the agricultural land from the Subcarpathian area where the slopes with reduced declivity are useful for agricultural exploitation of the soil (tillable—505.56 ha; complex cultures—140.43 ha).

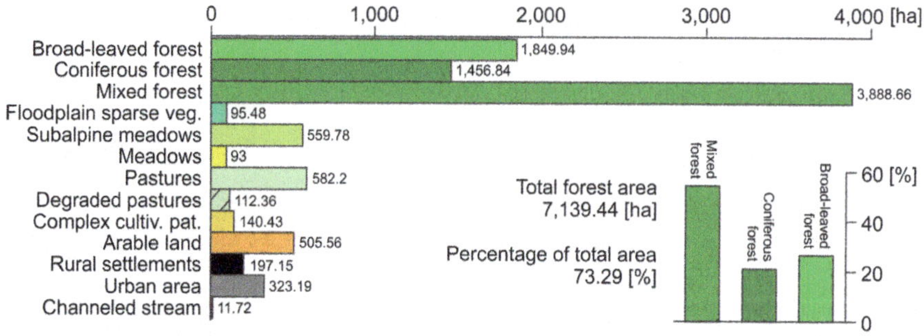

Fig. 4.25 Main types of land use within the Cuejdiu hydrographic basin in 2010

4.5.2 Soil Cover

Considering the vegetation, climate and interaction with the lithological substrate, all mentioned in the previous chapters, the soil from the studied region is divided in two distinct components: the soils from the mountain areas and the Subcarpathian hills, covered with tree vegetation, pastures or meadows from the Cuejdi Subcarpathian depression, where most part of the tillable land is located.

The soils from the forested area belong to two specific forestry types, namely Cambisols, amounting to 87% and Argiluvisols, 12.1%. In addition, there are areas with undeveloped soils forming on the surface of superficial deposits, totalling approximately 0.4% from the hydrographical basin. Using information taken from the "Forestry improvement of Gârcina Forestry District" journals divided into production units the soils from the forestry area are represented in UP Runcu by Cambisols, 97% and Argiluvisols, 3%, and in UP Mijlocul, where Cambisols amount to 71% from the total surface while the Argiluvisols to 29%. Regarding the soil subtypes, the most representatives are: the typical Eutricambisoil and the typical Districambisoil (Amenajamentele Silvice ale Comunei Gârcina 2005) (Fig. 4.26).

In the studied area on the Cuejdi Valley, nine taxonomical units were identified belonging to Protisoils, Cernisoils, Cambisols, Luvisoils, Hidrisoil, Antrisoil as well as 11 major subtypes represented at the map scale level. Following cartographic and statistical analyses the clear predominance of Cernisoils was observed, the other soil types having only a distinct spatial disposition (Studiu Pedologic–Teritoriul Comunei Gârcina 1998) (Fig. 4.27).

Protisoil class is represented by Regosols and Alluvial Soils. Regosols, frequently present on the steep slopes, covers a total surface of approximately 369.22 ha and are specific to the early stage of soil formation, being young soils, with hardly developed horizons. It frequently presents a short Ao horizon, followed by parental material; they are not fertile soils, with short horizons, due to gravel presence at shallow depth. The subtype met frequently is the Calcaric type.

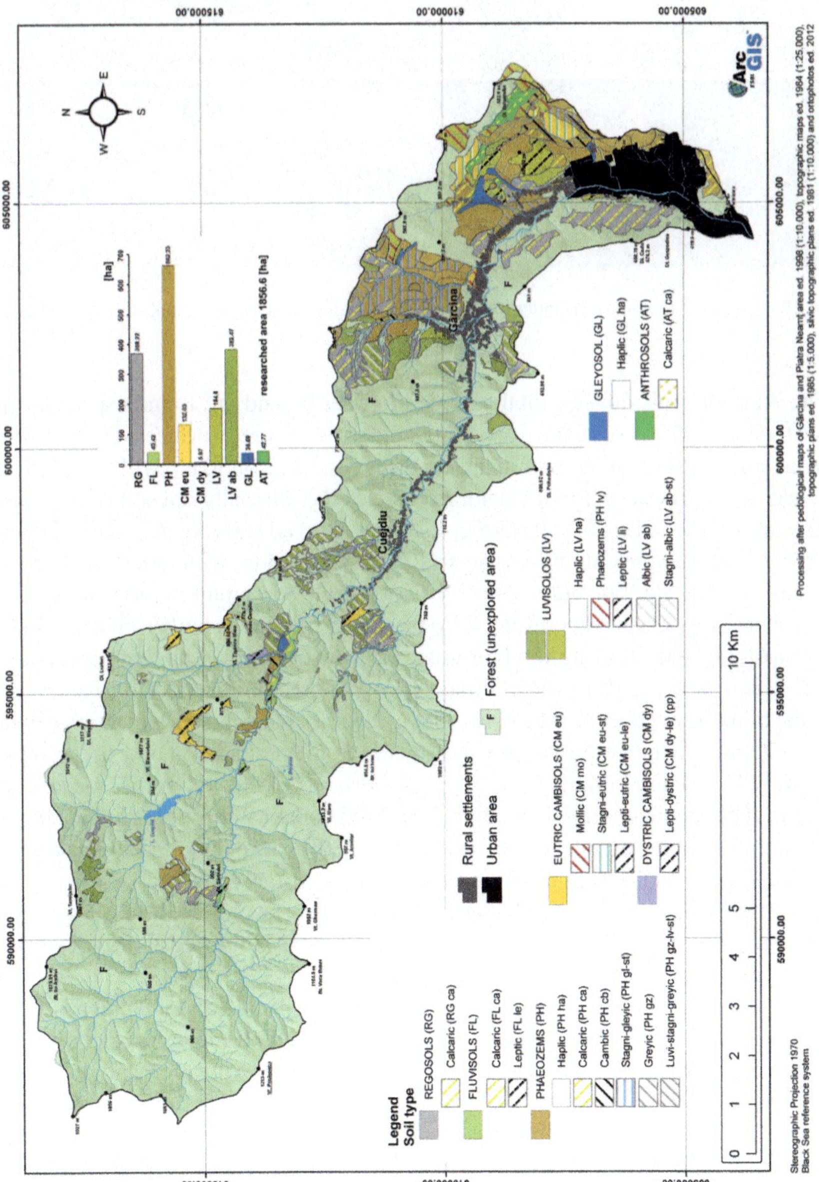

Fig. 4.26 Soil map of Cuejdiu hydrographic basin

Fig. 4.27 Layers of topsoil over underlying geological material on the left bank of Cuejdel Lake

The Alluvial soils cover a total surface of 40.42 ha and are found especially in the waterside sections of main rivers. Regarding the covered surfaces, the Calcaric Alluvial soils (ASka) are predominant, covering a surface of 30 ha, with the highest presence in the central sector of Cuejdi River; also the lithic alluvial soils (ASli) and the erosion-calcaric alluvial soils (AS er-ka). The Alluvial soils are generally defined by clayey-sandy texture, slightly or neutral alkaline reaction, with moderate amount of calcium carbonate. The humus reserve varies from very small to medium size.

The Cernisoils class is represented by faesioms, present in higher altitudes than the cernosioms and covering the largest surface (662.23 ha) of the cernisoil class, with frequent subtypes Calcaric, Cambic, Grey. The predominant faesioms are those from the clinogleyc subtype. They have clayey texture, with stagno-gleization (w) in the first 50 cm, the temporary water excess being from precipitation or coast springs, but mainly from lateral drainage, through the horizons of the soil profile located on the slope. The humus reserve is medium sized and contains phosporus and medium potassium. The saturation degree in bases is over 80%. The greyc faesioms present an Ame horizon, and the depth of the Cca horizon is usually 130 cm. The Calcaric faesiom represents a soil with carbonates from the surface or until the base of A horizon, until at most 50 cm and can be located in the South-Eastern part of the Cuejdi hydrographic basin.

The Cambisols class is represented by eutricambisoils and districambisoils. The Eutricambisoil (typical eumezobasic brown soil) formed on mineral rocks that are rich in calcium and iron on slopes with relatively diverse exposition and declivity, with a strong acid reaction (pH 4.5–7.1) and a high content of humus in the first 15 cm (7.2–9.3%). The soil texture ranges from clayey-sandy to clayey, favoring the evaluation of soil for spruce, fir, beech, due to the high edaphic volume and humidity regime and the favorable trophicity. In the Cuejdi basin it represents 7% from the total surface, respectively 132.03 ha, being especially specific to the upper basin; the most frequent subtypes are Molic, Stagnic and Lithic.

The Districambisols class is not conditioned by the exposition or slope, it has a strong acid reaction (pH 3.8–5.3) and a slight to moderate content of humus (3.9–7.2%) in the first 15 cm. The texture is glayey-sandy to sandy, favoring a superior evaluation for spruce, fir and beech. It develops especially at the Cuejdi springs, in the area of Pascău, Cerbul, Argintăriei and Dranişei creeks. The specific subtype in the Cuejdi hydrographic basin is the Lithic subtype and covers an area of 5.97 ha (0.32% from the total of studied area).

The Luvisoils class is represented by preluvisol and luvosol. The preluvisols cover limited areas in the South-Eastern part of the studied hydrographic basin (184.8 ha) and includes the Typic, Molic, Lithic subtypes. These soils have a differentiated texture in profile, glayey to the surface and glayey-loamy in the Bt horizon. These are soils with slight acid reaction and small humus reserve. The mobile phosporus content is small to medium and the potassium content is high and very high.

The Luvosoil formed on slopes with moderate declivity and diverse exposition, it has a strong acid reaction (pH 4.6–5.3), varied humus content (3.3–13.5%) and only 5–6 cm thick. The spreading of this soil type is specific for the medium-lower sector of the Cuejdi hydrographic basin and is more presnt in UP Mijlocu, between the Ciuculoaia and Gârceniţa creeks and in Mintiana, Balaurului and Turtureşti Hills and rare in UP Runcu.

The Hidrisoils class is represented by gleysoil. Th etypical gleysoil presents a geographical spreading focused on the shadier area of the intramontane space (36.69 ha), has an undifferentiated texture (from rough to fine) or contrasting, with undeveloped structure, unfavorable aerohydrical regime (strong surplus of water), less humus of low quality (2–3%). It is identified naturally on surfaces covered with low quality meadows.

The Anthrisoils class is represented by anthrosoil. The erosion-calcaric anthrosoil, with approximately 43 ha, is present on limited areas from the Eastern part of the territory, especially on steep slopes and in the vicinity of human settlements. These are very shallow soils (60–80 cm), with undeveloped structure, low content of humus and nutritional elements. It is strongly influenced by erosion with surface pedogenetical mineral horizons highly influenced by anthropic activities; they present a profoundly modified structure compared to the initial one that led to the presence of effervescence horizon to be located in the upper part of the profile (Studiu Pedologic–Teritoriul Comunei Gârcina 1998).

4.6 The Role of Anthropic Influence

The human-geographical database was realized based on the Romanian Statistical Directory (1957–2011) and Statistical Fascicles of Gârcina Commune, issued by the Regional Statistic Direction Neamț (1956–2010). Highly helpful were also the statistic data published in the monography of Gârcina commune by Baroi (2012).

The anthropic factor is the most dynamic elements in the hydrographic basin of Cuejdi River, being the result of human society actions to make capital out of local resources (forests, waters, etc.), contributing to the change of parameters from the natural environment. Maybe the most important aspect is represented by deforestation that leads to acceleration of erosion processes and degradation of exploited land.

In the studied area, forests represented the primary natural resource for population who settled in the Cuejdi Valley, a fact clearly demonstrated by the forest roads over 100 years old. Nevertheless, before the First World War, the access in the area that subsequently transformed into a natural dam lake was really difficult. First recorded action as such was made between 1910 and 1925, when a shareholder named Petre Juster, contracted for exploitation a surface of 425 ha of forest on the Vaca Babei creek. The access in the upper part of the hydrographic basin was realised by building a railway with a length of 7 km between Vaca Babei creek and the old timber factory located downstream, at the exit of mountain area. After 1925, the railway became a state property and entered into total degradation (Amenajamentele Silvice ale Comunei Gârcina 1963–2005).

After 1945, a new railway line for forest exploitation of the upper basin was built by the Romanian-Soviet society SOVROLEMN, linking the upper part of Cuejdi village to Mătieş–Bursucăria point (West of Cuejdel Valley). Finalized at the beginning of 1950s, it had a length of approximately 10 km and used as thrust for the wagons and a reduced tonnage locomotive specially designed for forest railways in the area. After only five years, it was extended until the factories from Piatra Neamţ, transporting daily between 15 and 20 wagons of wood. The railway functioned until the first part of 1960s, when it was considered that auto transport is more profitable and thus a road for forest exploitation was built over the old trail track of the railway line (Amenajamentele Silvice ale Comunei Gârcina 1963–2005).

Between 1967 and 1987, forest roads were built totalling 25.4 km, most important being: Prislop—2 km (1967), Putredu—1.5 km (1968), Cuejdi–Funduri—3.4 km (1971), Argintăria—2 km (1980), Prislop prelungirea—3 km (1985), Ţiganca—3 km (1985). From that period until present time, no other access roads were built in the upper basin of Cuejdi, the current forest exploitation being made using the old infrastructure.

A secondary route of forest exploitation that shortens the connection road between Cuejdi and Cracăul Negru represented one of the main factors in triggering the landslide that blocked the Cuejdel Valley. Initially, that road detached from the Cuejdel Valley to the right, before the confluence with Cuejdel creek and followed it until the confluence with Glodu creek, crossing the bypass road again. On this route, the forest road was intensely used until 1991 and sectioned the base of the slope under the Muncelul Peak triggering large gravitational displacements over approximately 1 km length (Fig. 4.28).

After the natural dam lake was formed, the road became unusable for heavy tonnage vehicles due to the section being buried under a huge deluvial mass. Currently, over the surface of the landslide, a series of access roads were rebuilt; the roads are only used by light vehicles (carts, etc.).

The influence of anthropic factor in the formation and evolution of the Cuejdel lacustrine basin had and still has an essential role. Destabilising the slopes through deforestation of large areas and the building of improper forest transport roads are

Fig. 4.28 Forest road network on the left bank of Cuejdel lake

directly responsible for the future dynamics of the lake. As long as there are no interventions to the slopes surrounding the lake or at the catchment level, the lacustrine basin will evolve into a normal regime.

References

Apăvăloae M, Apostol L, Pîrvulescu I (2001) Caracteristici ale umezelii aerului din Subcarpații Moldovei. Luc Sem Geogr D Cantemir—Univ Alexandru Ioan Cuza, Iași 19:227–235

Apăvăloae M, Apostol L (2002) Durata de strălucire a soarelui în Subcarpații Moldovei. Luc Stat Stejaru, S Geogr 11 (Piatra Neamț)

Apostol L (2004) Clima subcarpaților moldovei. Suceava University Publishing, Suceava

Atanasiu I (1939) Contributions à la stratigraphie et la tectonique du Flysch marginal Moldave. Ann Sci Univ Jassy 25 (Bucharest)

Atanasiu I (1943) Les faciés du Flysch marginal dans la partie moyenne des Carpathes Moldaves. Ann Inst Geol Rom 22 (Bucharest)

Athanasiu S (1905) Asupra stratigrafiei Muntelui Stînişoara. Bull Soc Sci Ann 14 (Bucharest)

Athanasiu S, Macovei G, Atanasiu I (1927) La zone marginale du Flysch dans la partie inferiéure du bassin de la Bistriţa. Assoc Géol Carp, II-éme réunion, Guides des excursion, Bucharest

Baroi C (2012) Gârcina, pagini de monografie (Gârcina, monographic pages). Cetatea Doamnei Publishing House, Piatra Neamţ

Băncilă I (1955) Paleogenul zonei mediane a flişului. Bull Sci Acad Rom 7(4) (Bucharest)

Băncilă I (1958) Geologia Carpaţilor Orientali. Scientific Publishing, Bucharest

Băncilă I, Papiu VC (1962) Asupra litologiei sedimentelor cretacice din anticlinalul Cârnu–Valea Ţiganilor (Regiunea Bicaz) I. Complexul inferior; II. Complexul superior, DS Inst Geol 45 (Bucharest)

Băncilă I, Hristescu E (1963) Linia externă şi linia pericarpatică dintre valea Sucevei şi valea Trotuşului. Asoc Geol Carp–Balk 5, Editura Tectonică, Bucharest

Botez C, Nichita O, Ionesi I, Iorga N, Căruntu C (1967) Studiul geochimic comparativ între stratele de Biserican de pe valea Bistriţei şi cele de pe valea Humorului. Stud Cerc Geol Geofiz Geogr–S Geol 12(1) (Bucharest)
Dumitrescu I (1952) Studiu geologic al regiunii dintre Oituz şi Coza. Ann Com Geol 24 (Bucharest)
Filimon T, Damian A (1965) Geologia regiunii Bicaz-Piatra Neamţ. SSNG Com Geol 3 (Bucharest)
Filipescu MG (1936) Recherches géologiques entre la valée du Teleajen et la vallée de la Doftana. Ann Inst Geol Rom 17 (Bucharest)
Filipescu MG, Draghinda I, Mutihac V (1962) Contribuţii la orizontarea şi stabilirea vârstei şisturilor negre din zona mediană a flişului Carpaţilor Orientali. Com Acad RP Rom 1 (Bucharest)
Gâştescu P (1963) Lacurile din Republica Populară România–geneză şi regim hidrologic. Romanian Academy Publishing House, Bucharest
Grasu C, Catană C, Grinea D (1981) Quelques considérations pétrogénétiques sur les dysodiles oligocénes des Carpathes Orientales Roumaines. Anal Sci Univ Alexandru Ioan Cuza Iaşi, S Geol–Geogr 27 (Iaşi)
Grasu C, Catană C, Grinea D (1988) Flişul carpatic–Petrografie şi consideraţii economice. Editura Tehnică, Bucharest
Ichim I (1969) Contribuţii la studiul geomorfologic al versanţilor din sud-estul Munţilor Stânişoarei. Lucr Stat Cerc Stejarul 2 (Pângăraţi-Neamţ)
Ichim I (1972) Văile elementare din Munții Stânișoarei și raportul lor cu structura și litologia. Stud Cerc Geol Geofiz Geogr, S Geogr 19(2):175–182
Ichim I (1973) Rôle de la structure et de la lithologie dans la morphologie des vallées des Monts Stânişoarei. Stud Geomorph Carp–Balc 7 (Bucharest)
Ichim I (1979) Munţii Stânişoarei–studiu geomorfologic. Romanian Academy Publishing House, Bucharest
Ichim I, Rădoane N, Rădoane M (1996) Procese geomorfologice cu intervale de recurenţă mare în arealul munţilor flişului. Exemplificări din Judeţul Neamţ (Geomorphological process with high frequences in flysch mountains area. Examples from Neamț County). In: Studii şi cercetări—Extras 8, Muzeul de Ştiinţe Naturale Publishing, Piatra-Neamţ, p 15
Idriceanu T, Ionesi L, Iorga N, Erhan V (1967) Contribuţii la studiul mineralogic şi sedimentologic al argilelor roşii din zona şisturilor negre. SSNG Com Geol 4 (Bucharest)
Ionesi L (1971) Flişul paleogen din bazinul văii Moldovei. Romanian Academy Publishing House, Bucharest
Ionesi L (1982) Asupra limitei Eocen-Oligocene în flişul extern carpatic. Luc Sci Univ Alexandru Ioan Cuza Iaşi–Cobălcescu G (Iaşi)
Ionesi L, Tocorjescu M (1968) Date microfaunistice asupra limitei Cretacic superior-Paleogen în flişul extern din bazinul văii Moldovei. Anal Sci Univ Alexandru Ioan Cuza Iaşi, S Geol–Geogr 14 (Iaşi)
Ionesi L, Grasu C (1976) Asupra prezenţei unităţi Vrancea între Suha Mare şi Nemţişor. Ann Muz St Nat Piatra Neamţ, S Geol–Geogr 3 (Piatra Neamţ)
Ionesi L, Gheaţă N (1978) Vârsta stratelor de Gura Șoimului din semifereastra Humorului. Ann Muz St Nat Piatra Neamţ, S Geol–Geogr 4 (Piatra Neamţ)
Ionesi L, Bogatu L (1986) Contribuţii asupra litostratigrafiei şi biostratigrafiei Formaţiunii de Gura Șoimului din semifereastra Bistriţa. Ann Muz St Nat Piatra Neamţ, S Geol–Geogr 5 (Piatra Neamţ)
Joja T (1952) Structura geologică a flişului marginal din regiunea Suha Mică şi Suha Mare. DS Inst Geol 36 (Bucharest)
Joja T, Chiriac M (1964) Asupra prezenţei unor amoniţi în Stratele de Hangu din flişul extern de la Putna. Stud Cerc Geol Geofiz Geogr, S Geol 9(1) (Bucharest)
Joja T, Mirăuţă E, Alexandrescu G (1968) Harta geologică 1:200000, L-35-IX, No 13, Piatra Neamţ–Notă explicativă, Inst Geol (Bucharest)
Joja T, Manoliu E (1970) Sur le présence d'une faune d'ammonites dans les Schistes noirs de l'anticlinal Horaița–Doamna, dans le ruisseau Sarata aux environs de la ville de Piatra Neamț. Rev Rom Geol Geophis Géogr, S Géol 14(2) (Bucharest)

Macovei G, Atanasiu I (1925) Structure géologique de la vallée de la Bistriţa. CR Inst Géol Roum 8 (Bucharest)

Macovei G, Athanasiu S, Atanasiu I, (1927) La zone marginale du Flysch dans la partie inferiéure du bassin de la Bistriţa. Assoc Géol Carp, II-éme réunion, Guide des excursion, Bucharest

Matei V, Brustur T, Grinea D, Stamate I, Catană C (1986) Nisipuri cuarţoase miocene din zona Calu–Tazlău—Consideraţii geochimice şi tehnico-economice. Ann Muz Nat Piatra Neamţ, S Geol–Geogr 5 (Piatra Neamţ)

Micu M (1979) Nouvelles données sur la stratigraphie et la tectonique du flysch externe du bassin de la Suceviţa. DS Inst Geol Geofiz 46 (Bucharest)

Mihu–Pintilie A, Romanescu G (2011) Morphometric and morphological suitability of the relief from the Crucii (Cuejdel) Lake basin (Stânişoarei Mountains). In: Pandi G, Moldovan F (eds) Proceedings of the air and water components of the environment, vol 1. Babeș–Bolyai University, Cluj, 18–19 March 2011, pp 305–313

Mihu-Pintilie A, Romanescu G, Stoleriu C (2012) Morpho–bathymetric parameters of recess Crucii (Cuejdel) Lake (Stânişoarei Mountains). In: Pandi G, Moldovan F (eds) Proceedings of the air and water components of the environment, vol 1. Babeș–Bolyai University, Cluj, 23–24 March 2012, pp 445–452

Mihu-Pintilie A, Romanescu G, Stoleriu CC, Nicu IC, Asăndulesei A, Schmaltz E (2014) Natural dam lakes from Cuejdiu watershed (Stânișoarei Mountains)—Non–invasive methods used for bathymetric maps. In: Gâştescu P, Marszelewski W, Breţcan P (eds) Proceedings conference of the water resources and wetlands, vol 1. Tulcea, 11–13 September 2014, pp 130–137

Minea I, Romanescu G (2007) Hidrologia mediilor continentale–Aplicaţii practice. Demiurg Publishing House, Iaşi

Mirăuţă O (1962) Stilul tectonic al flişului marginal şi al molasei subcarpatice din regiunea Piatra Neamţ. DS Com Geol 48 (Bucharest)

Mirăuţă O (1969) Stratigrafia şi structura Miocenului subcarpatic din regiunea Moineşti–Tazlău. DS Inst Geol 54(3) (Bucharest)

Mirăuţă O, Mirăuţă E (1964) Stratigrafia flişului cretacic şi paleogen din valea Cuiejdului şi valea Horaiţei. DS Com Geol 5(1) (Bucharest)

Mutihac V, Ionesi L (1974) Geologia României. Editura Tehnică, Bucharest

Olaru L (1978) Cercetări asupra distribuţiei stratigrafice a microflorei în flişul palegen dintre văile Bistriţei şi Trotuş. Mem Inst Geol–Geofiz 27 (Bucharest)

Olteanu C (1952) Cercetări geologice între valea Bistriţei, pârâul Cuiejdiu şi pârâul Pângărăcior. DS Inst Geol Rom 36 (Bucharest)

Olteanu C (1953) Revizuiri geologice la sud de Piatra Neamţ. DS Inst Geol Rom 37 (Bucharest)

Papiu VC (1960) Petrografia rocilor sedimentare. Editura Științifică, Bucharest

Papiu VC, Alexandrescu G, Ţăranu E (1967) Studiul mineralelor grele din seria argilelor vărgate din pânza şisturilor negre. DS Inst Geol Geofiz 62(1) (Bucharest)

Papiu VC, Alexandrescu G, Iosof V (1983) Considérations sur le genèse des schistes argileux bigarrés (Vraconien-Turonien) du flysch des Carpathes Orientales de Roumanie. Ann Inst Geol Geofiz Carp–Balk 62 (Bucharest)

Polonic P, Polonic G (1968) Direcţii de transport în molasa helveţiană dintre valea Sucevei şi valea Trotuşului. DS Com Geol 54(1) (Bucharest)

Rădoane N (2002) Un nou lac de baraj natural în bazinul Bistriţei Moldoveneşti-Lacul Cuejdel. Stud Cerc Geograph 49:76–82

Romanescu G (2000) Limnologie şi Glaciologie-Partea I Limnologie. Suceava University Publishing, Suceava

Romanescu G (2009a) Trophicity of lacustrine wetlands on the Carpathian territory of Romania. a case study from the East Carpathian mountains. Luc Sem Geogr D Cantemir, Univ Alexandru Ioan Cuza Iași 29:5–13

Romanescu G (2009b) Trophicity of lacustrine waters (lacustrine wetlands) on the territory of Romania. Lakes–Reservoirs Ponds—Rom J Limn 3:62–72

Romanescu G (2009c) The physical and chemical characteristics of the lake wetlands in the central group of the east Carpathian Mountains. Lakes Reservoirs Ponds—Rom J Limn 4:94–108

Sandu I, Pescaru VI, Poiană I (2008) Clima României. Romanian Academy Publishing House, Bucharest

Săndulescu M, Săndulescu J (1964) Aspecte stratigrafice şi structurale ale flişului paleogen din regiunea Ghelinţa. DS Com Geol Rom 49(1) (Bucharest)

Stoica C (1944) Paleogenul din zona Sibiciului. Rev Muz Geol–Min 8(1) (Cluj)

Stoica C (1953) Stratele de Gura Șoimului. Anal Univ CI Parhon 2 (Bucharest)

Ştefan P (1974) Asupra prezenţei lui Inoceramus tuberculatus în Stratele cu inocerami din flişul extern. Anal Sci Univ Alexandru Ioan Cuza Iaşi, S Geol–Geogr 20 (Iaşi)

Turculeţ I (1971) Inoceramus inconstans Woods em Studert, în Stratele cu inocerami de la Pângăraţi (unitatea de Tarcău). Luc Stat Stejaru, S Geol–Geogr 4 (Piatra-Neamţ)

Vinogradov C, Pârvu G, Bomboe P, Negoiţă V (1983) Petrografia aplicată a rocilor detritice. Romanian Academy Publishing House, Bucharest

Zăvoioanu I (1978) Morfometria bazinelor hidrografice. Romanian Academy Publishing House, Bucharest

Chapter 5
Genesis of the Cuejdel Lake and the Evolution of the Morphometric and Morpho-Bathymetric Parameters

Abstract The present lacustrine basin of Cuejdel Lake has been forming in the upper basin of Cuejdiu river at about 1.5 km upstream Cuejdel brook. Its origin is based on a large landslide that produced the rock displacement during the summer of 1991. The morphometric parameters for landslide were extracted from digital terrain model (DTM) and orthophoto plans (Ed. 2012, 2010 and 2006) using GIS software (ArcGIS, Global Mapper and TNTMips). In order to determine the morpho-bathymetric parameters of the lake basin, direct methods were used: Dual Frequency Echo Sounder for bathymetric survey and Leica System 1200 (TCR 1201 and GPS 1200) for aquatic surfaces and shorelines. From measurements made over the entire surface of the lake between 2011 and 2014, approx. 45,000 bathymetric levels resulted, on which several digital elevation models were conducted of the lacustrine depression with 0.25 m/pixel resolution (DEM 2011 and DEM 2014). The bathymetric maps results were used to determine the morphometric (aquatic surface or water mirror surface, length and width of the lake, lake major axis and minor axis, and perimeter or length of shoreline) and bathymetric parameters (actual shape of lacustrine basin, average and maximum depth, average slope of the lake bottom, water volume expressed as the total or partial volumes).

The emergence of the natural dam lake Cuejdel within the landscape of the Stânişoarei Mountains (Eastern Carpathians) was originally reported in several scientific articles by researchers from the Research Station "Stejarul" in Piatra Neamț, the first and most detailed being published by Ichim et al. (1996). From here we were able to take the first details concerning its location, some summary data on the genesis and the evolution of the water surface, or on the indicative variation of morphometric parameters (such as maximum depth or water level elevation).

At the beginning of this study, the only available cartographic database was the bathymetric sketch executed in 1992 by Ioniță and his associates, and retouched by Rădoane (2002). Although the morphometric values issued by both authors were obtained using empirical methods, the bathymetric model proved itself useful during their investigations. Later, official data on the lacustrine unit were few, and often proved inaccurate, which prompted the creation of the current bathymetric support,

A. Mihu-Pintilie, *Natural Dam Lake Cuejdel in the Stânişoarei Mountains, Eastern Carpathians*, https://doi.org/10.1007/978-3-319-77213-4_5

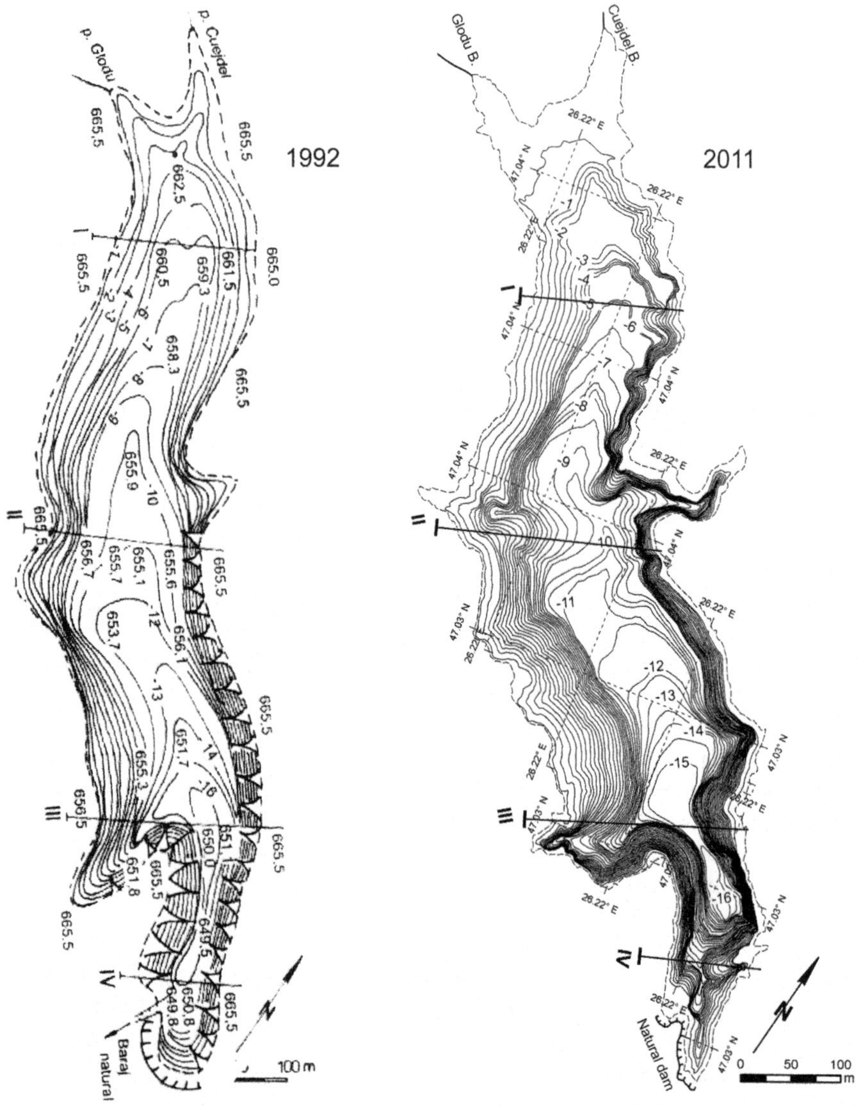

Fig. 5.1 Bathymetric comparisons of lacustrine basin between 1992 (after Ichim et al. 1996; Rădoane 2002) and 2011 (Mihu-Pintilie et al. 2012, 2014a)

using a high precision technical toolkit. The measurement method is explained in more detail in the next section (Fig. 5.1).

5.1 Cuejdel Lake Formation Stages

The Cuejdel stream (length—3.79 km) is the most important tributary of Cuejdiu river (Order IV—Horton-Strahler classification), from which it drains an area of 10.38 km^2 (10.64%). Its valley is oriented north-south and generates a series of secondary valleys that originate almost transversely from the main stream (Piciorul Crucii and Glodu—right slope; Hălăgeanu and Făgetu—left slope). Due to its confluence with the Glodu creek, the watershed took a bilobate shape, inducing a circularity coefficient of the catchment of 0.67 (Mihu-Pintilie and Romanescu 2011).

The slopes are characterized by high declivity (15°–45°); the interfluvial heights frequently exceed 1000 m (to the west, Mătieş Peak—986 m; to northwest, Tarniţelor Peak—1081 m; to northeast, Măgura Hill—1117 m; to the east, Muncelul Peak—1077 m), and the valleys describe a transverse profile slightly larger in the upper, and narrower in its lower part (Ichim 1972, 1973). Prior to the landslide that blocked the stream in the 0.5–1.5 km upstream of the confluence with Cuejdiu river, the valley originally dug a small gorge into the permissive lithology. Currently, the creek has resumed erosion, deepening its layer at the contact between the sliding mass and the right side of the valley (Mihu-Pintilie et al. 2014a).

In this geomorphological context, gravitational tendencies to bar the valley took place for the first time in the summer of 1978, when it formed a smaller lake. The causes that led to the manifestation of the slope processes came under the friable lithology of the superficial deposits, driven by the erosion activity of the Cuejdel stream. The right side of the valley, destabilized also by the logging road connecting the valley of Cuejdiu and the valley of Cracăul Negru, dislocated in an area of approx. 300 m long, the disjointed materials occupying the entire riverbed. The scarp was reinstated one year later, in the synoptic context of June and August, when the maximum rainfall at all the weather stations in the area was recorded. Judging also by the historical discharge data recorded in 1979 on Cuejdiu River (45.2 m^2/s—22.VI; 77.2 m^2/s—15.VIII), the flow activity in the upper basin was intense, causing significant changes in the local landscape (Ichim et al. 1996).

The water accumulation from that year was estimated by various authors to have 200–250 m length and 4–5 m max. depth (Ichim et al. 1996; Rusu et al. 2002; Rădoane 2002). However, the forest plans made in 1981 (scale of 1:10,000) indicate that the lake was mapped with much larger dimensions: surface—2.29 ha, maximum length—451.8 m, average width—50.2 m. Based on the reconstruction of the flooded valley and on the deduction of the lake basin morpho parameters, the real average depth was 2 m, and the maximum depth of about 5–5.5 m. According to the reconstituted bathymetric model, the total volume of water collected behind the obstructive dam was about 45,000 m^3, rising progressively up to resume the flow on the deluvial cover. The maximum level was maintained until 1986, when several floods have removed much of the dam, the accumulation been almost completely drained (Mihu-Pintilie et al. 2014a; Stoleriu et al. 2014).

After a series of events that substantially changed the configuration of the Cuejdel Valley, which led to immersing or flooding of 0.5 km of forest road, the traffic induced by logging in the area resumed erratically on several tracks, routed to the middle part of the slope under the Muncelul Peak. In 1985, on an area of only 46 hectares, located between Făgetu brook and the confluence of Cuejdel with Cuejdiu, there was a density of road network of 10.67 km/km^2, in the context of a 30°–50° slope and of a relief intensity >300 m. Basically, the slope fragmentation by anthropic intervention was one of the preconditions for the manifestation of the further extensive slope processes.

The emergence of the actual lake has occurred in the summer of 1991, when the landslide was reactivated on a much larger scale as previous (Ichim et al. 1996). The main morphological landmarks of the landslide can be summarized as follows: maximum altitude of the detachment niche—930 m (left lobe), 820 m (right lobe); maximum fall—349 m; landslide length—1062 m; blocked stream length—1215 m; sliding affected area—67.74 ha; deluvial mass depth—5.0–25.0 m. The main factors that led to the slope imbalance were: reactivation of the old landslide produced during 1978 and of some lithological detachments located at the top of the slopes; heavy rains in May and June (741.4 mm); the density of the forest roads network (10 km/km^2); the 5.4° earthquake on Richter scale in 1990 (Ichim et al. 1996; Rusu et al. 2002; Mihu-Pintilie et al. 2014a, b, c, d) (Fig. 5.2).

The landslide complex resulting from the rotational movements of hard rocks displaced from the top of the slopes, combined with the mudflows and solifluctions directed to torrential valleys gave birth to a chaotic relief, set in catsteps, earth waves, mounds and micro-depressions where water stagnates temporarily. The scarps formed in situ in rock hold heights of between 10 and 15 m, being very active even today. Ruptures formed on the deluvial surface have a lower magnitude, and their activity is diminished by the forest vegetation. However, torrents installed at the contact between the earth waves and the lithological resistance remnants maintain, in some sectors, abruptions of the friable clay-marl soil mass (Fig. 5.3).

The violent character of the landslide affected the forest too, through tree falling or root instability, in some areas being preserved with the appearance of "drunken forest". Also, part of the infrastructure fitting the operating road at the base of the slope (bridges, concrete poles, drains) was buried under the deluvial mass. Based on the mapping of existing operational roads in 1985, the result was that the deluvial mass covered 1.91 km of main road and 4.08 km of secondary trails.

The Cuejdi creek bed was sunken within a very short time. The way of filling with obstructive materials and their depth divide the stuck stream in two distinct sectors. The first section, 396 m long, lies from the lake limit to the downstream accumulation, known as Constellation Lake. Here, the deluvial thickness registers the highest value (>25 m), overlapping the old dam formed 12 years ago (in 1979).

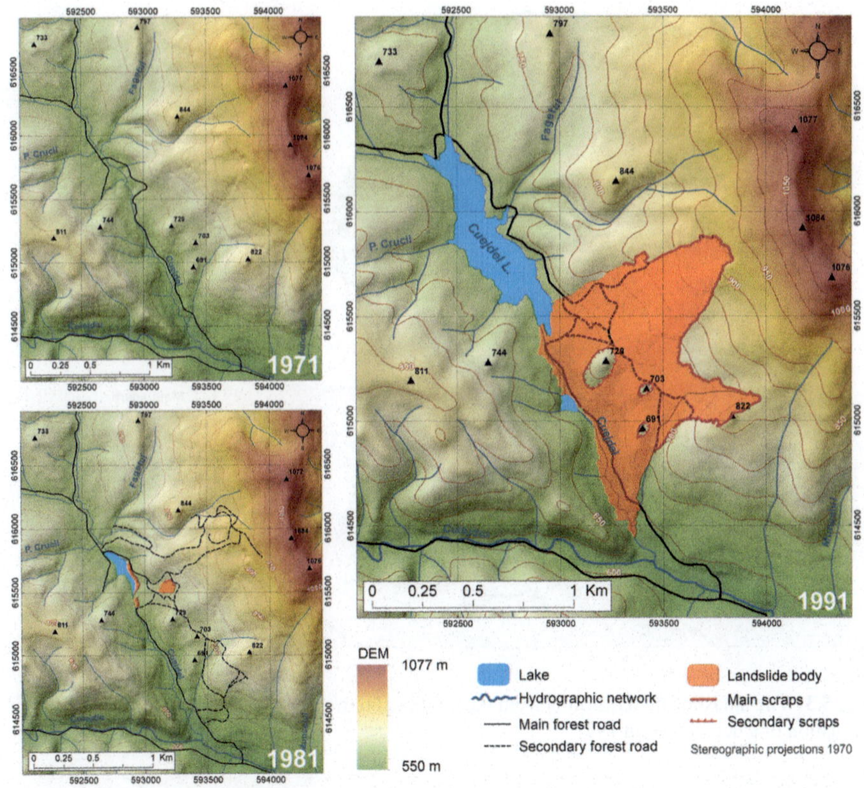

Fig. 5.2 The evolution of landslide under Muncelului ridge and lake Cuejdel formation stage using cartographic data. processing after: silvic topographic plans ed. 1971 and 1981 [1:10,000], topographic maps ed. 1984 [1:25,000], topographic plans ed. 1985 [1:5,000] and ortophotos ed. 2012

The narrow transverse profile (15–20 m) and the sizable thickness of the valley (20–30 m) made the dam have bigger amplitude relative to its width. The high density of deposits carried by the left landslide corridor, made of cuts of massive sandstone detached from the main scarps and caught into a more robust clay matrix, induced a very high degree of compaction. Even if the landslide blocked >1 km in length of Cuejdel Valley, the current dam lake is just over 300 m in length (Fig. 5.4).

The second sector is characterized by a thinner dam, due to the widening of the riverbed, but also to the high homogeneity of materials deposited on the valley floor. The finer fraction of the deluvial deposits is due to the more clayey source detached from the south-western slope of Muncelul Peak. Also, the lithological resistance remnants stemming the landslide channeled the deluvial mass movement in several corridors, determining a non-uniform filling of the riverbed. Thus, downstream two small lakes were formed with a cascade disposal behind the secondary dams (Con-

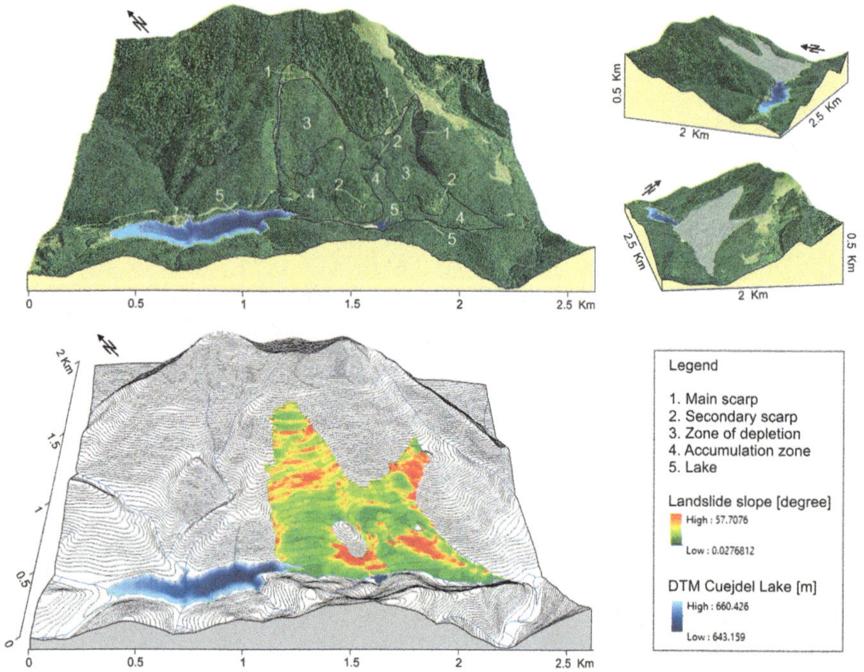

Fig. 5.3 The landslide dam resulting from the rotational movements of hard rocks and mechanism formation of Cuejdel lake in 1991

Fig. 5.4 Current landforms and geomorphological processes in the dam area of Cuejdel lake

Fig. 5.5 Lake basins formed along the Cuejdel valley by landslide dam: **a** Cuejdel lake; **b** Constellation lake; **c** Blue lake

stellation Lake, Blue Lake) (Rusu et al. 2002; Mihu-Pintilie and Romanescu 2011) (Fig. 5.5).

Cuejdel Lake experienced a rapid increase in water level due to the particularities of the flooded valley sector and to the high amplitude of the dam. "S"-shaped, the lake keeps the configuration of Cuejdel stream, from which it took the main features: north-south flow direction, banks festoon next to major tributaries (Piciorul Rotund, Crucii and Făgetu stream), a short and steep left slope, and a less deep right bank, covered with forest vegetation.

According to the information taken from the topographical plans (1:5000) made in the year 1985, the water accumulation flooded 8.21 hectares of forest, mostly made of spruce, of which 5.7 ha were aged >20 years, and the remaining 2.51 ha were underbrush. Based on the orthophotoplans created from the aerial photos taken in 2012, it has been mapped a number of approx. 3358 trees still standing upright and emerging with approx. 0.5–5 m above the water surface.

Regarding the transport infrastructure, no less than 1.48 km of logging roads and four bridges were submerged. Currently, much of the road was reshaped on the left side of the lake (length—1.18 km) and only one bridge was restored over the Cuejdel stream, approx. 100 m upstream from the lake. However, despite the efforts of local authorities to relieve the passage between the valley of Cuejdiu and the valley of Cracăul Negru through the old route along the Cuejdel brook, currently this sector can be covered only by foot or light carts. The landslide under Muncelul Peak is not unique within the landscape of Stânişoarei Mountains, the frequency and complexity of gravitational movements of this kind being so numerous in this Carpathian sector and other units constituted on flysch deposits. Examples of landslides with medium and high recurrence interval which reactivate old deluvia, more or less stabilized, are multiple; enlightening in this regard are those from Buzău Valley, Putna basin (Vrancea County), Bistrița Valley (Grigore and Ielenicz 1972; Surdeanu 1986). The particularity of the landslide from the Cuejdel Valley is given by favorable conditions to support the emergence of a large scale natural dam lake without affecting riparian human society, and without causing significant damage (Ichim et al. 1996; Mihu-Pintilie and Romanescu 2011; Mihu-Pintilie et al. 2012).

5.2 Dynamics of Morphometric and Morpho-Bathymetric Parameters

The main morphometric parameters of a lake are related to the following plan variables: aquatic surface or water mirror surface, length and width of the lake, lake major axis and minor axis, and perimeter or length of shoreline. Morpho-bathymetric parameters indicate the form of the lake basin: actual size of lake basin, average depth, maximum depth, average slope of the lake bottom, water volume expressed as the total or partial volumes. Also, in addition to these were added other coefficients calculated based on geometric formulas that combine the two types of indicators: coefficient of shore sinuosity, form coefficient or axle ratio, shape index of the lacustrine depression (Kalff 2003; Minea and Romanescu 2007; Romanescu et al. 2013).

The method of mapping Cuejdel lacustrine basin was based on echo-sounding technology, currently used for creating maps which target the submerged coastal morphology from the lacustrine units or riverbeds. In this case, it has been used a high precision echo-sounder with dual frequency of acoustic signal (210/33 kHz) and a GPS system. From measurements made over the entire surface of the lake in consecutive campaigns between 2011 and 2014, resulted approx. 45,000 bathymetric levels, on which there were conducted several digital elevation models of the lacustrine depression with 0.25 m/pixel resolution (DEM 2011 and DEM 2014) (Mihu-Pintilie et al. 2012, 2014a, b, c, d).

Using the bathymetric maps made in 2011 and 2014, and the bathymetric sketches performed by Ichim and collaborators in 1992, there have been extracted the morphometric and bathymetric parameters that characterized the lake basin in various stages of evolution. The database was supplemented by extracting planimetric information from orthophotoplans (ed. 2005, 2010, 2012) and by making DEMs based on topographic plans (scale 1:10,000—ed. 1981; 1:5,000—ed. 1984). These are able to capture the topographic reality in a detailed manner before the valley is flooded. Also, some information in the literature about hydro-geomorphological events which brought significant changes into the lake structure were introduced within the diachronic analysis, but not before their validation on site, both by mea-

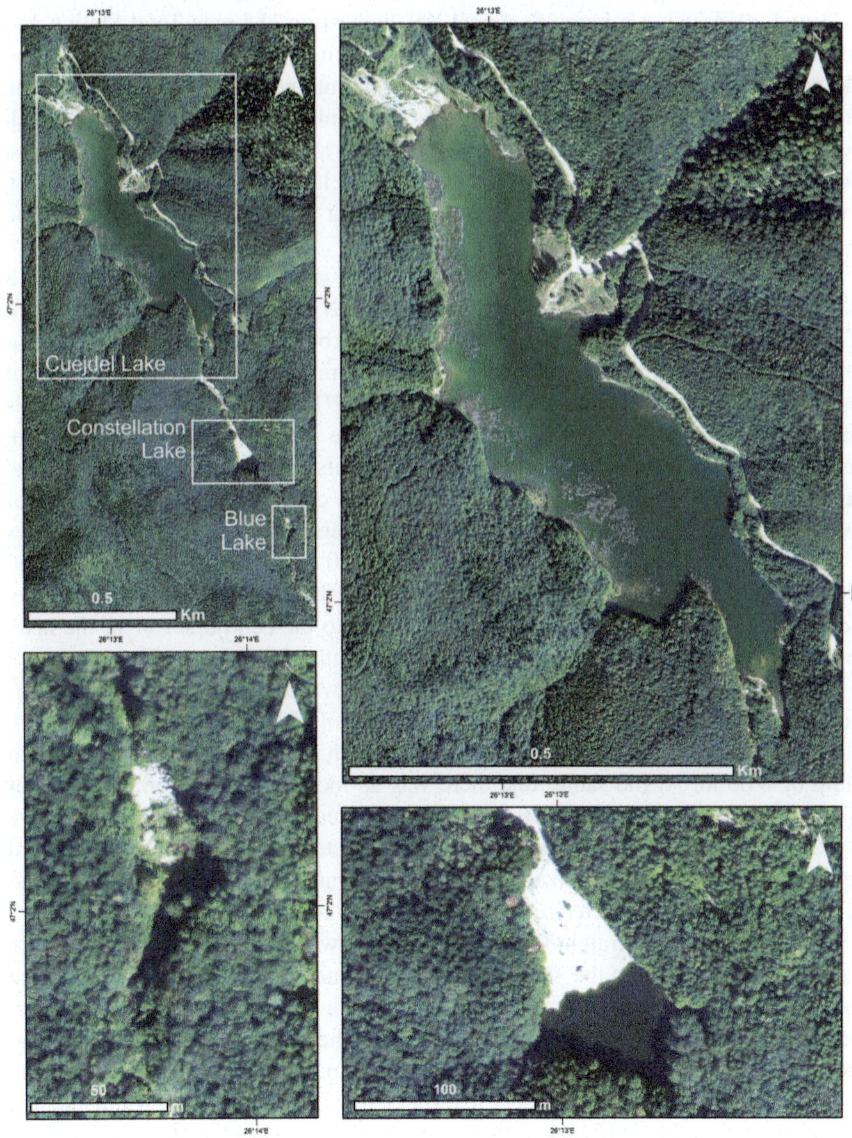

Fig. 5.6 Aerial image of Cuejdel lake in 2013 (orthophoto, www.ancpi.ro)

suring the level variations indicators from the trees trapped into the water mass, and by determining the deepening rate of the drainage channel (Fig. 5.6).

A lake surface is the area occupied by the water mirror at a time, and is influenced by the water level and by the season in which the measurements occur. In the case of Cuejdel Lake, the aquatic surface was determined through field measurements in

2011 and 2014, using a Leica System 1200, consisting of Leica Total Station and Leica TCR 1201 GPS 1200, by which it has been done the topographic survey of the lake contour. In 1991, the original surface was determined by retracing the shoreline based on the level indicators from trees in the proximity of the current lake, to which were added information taken from the bathymetric sketch made in 1992. Thus, the maximum surface of the lake was about 16.22 ha, indicating hitherto a yearly average loss of the water mirror of about 0.1 ha/year (1067 m^2/year) (Mihu-Pintilie et al. 2012) (Fig. 5.7).

In 2011, the surface occupied by the water mirror was 13.95 hectares, and in 2013 (13.88 ha) the area diminished in just two years with only 0.07 ha. This impairment of the surface, far below the multiannual ratio, can be explained by the low hydro-geomorphological dynamics of the lake in recent years. Basically, the most significant losses occurred in the first stage of its existence (1991–1992) due to repeated floods that have deepened the drainage channel from the surface of the landslide dam. Subsequently, the waterway has gradually changed its course closer to the right side, situated approx. −2 m lower than the initial outlet, causing a dramatic decrease of levels, and implicitly of the morphometric parameters (Fig. 5.8).

The maximum length of the water body was 1170 m (1.17 km) and was recorded in 1991, when the drain restarted on the surface of the deluvial mass. At that time, the maximum width was 341.2 m, corresponding to the central sector of the lake, and the segment between the right side and the mouth of Făgetu creek.

The average width, determined by the ratio between the surface and the maximum length, stood around 138.6 m. Measurements of 2011 (length—1004 m; maxi. width—282.6 m) revealed an impairment of length of −166 m and of maximum width of −60 m. At the same time there was an increase of the average width of +0.2 m (average width—138.8 m) due to the significant loss of surface area relative to the overall length. Two years later, these parameters showed a slight downward trend, but it must be taken into account that the negative fluctuation was intensified by the dry spell that characterized the period of 2012–2013.

The major axis is a straight line connecting the two most distant points of a lake even though it intersects the shoreline, and the minor axis is perpendicular to the longitudinal axis, connecting two points on opposite banks at the greatest distance. Immediately after the formation of Cuejdel Lake, the major axis was 1154 m long, and the minor axis was 358.4 m, and in 2013 the major axis was 956.9 m and the minor axis 285.3 m. Values very close to those of maximum length and width indicate a form developed almost linearly following the flooded riverbed, without side valleys to greatly influence these two parameters. Based on the ratio between the major axis and the minor axis of an irregular plane form it was determined the evolution of the form coefficient of the lake (1991—0.31; 2011—0.30; 2013—0.29). The results revealed an oblong shape specific to natural dam lakes, driven by the much greater length in relation to the average width (ratio approx. 3/1). This elongation trend of the surface occupied by the water mirror is directly proportional to the mass of liquid loss, validating the form of the first lacustrine accumulation, which in 1981 had an axle ratio value of 0.25 (ratio approx. 4/1).

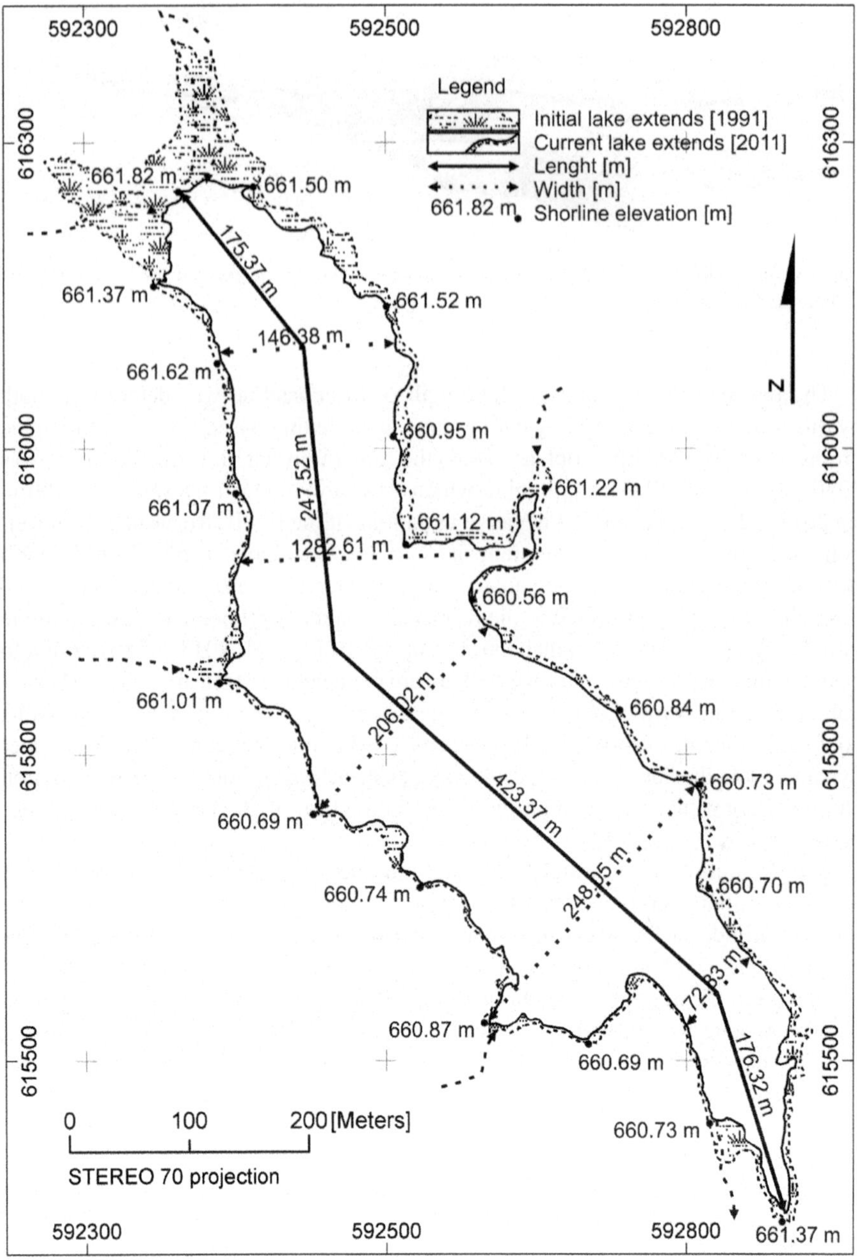

Fig. 5.7 The topographical survey of the Cuejdel lake in 2011 using the Leica system 1200

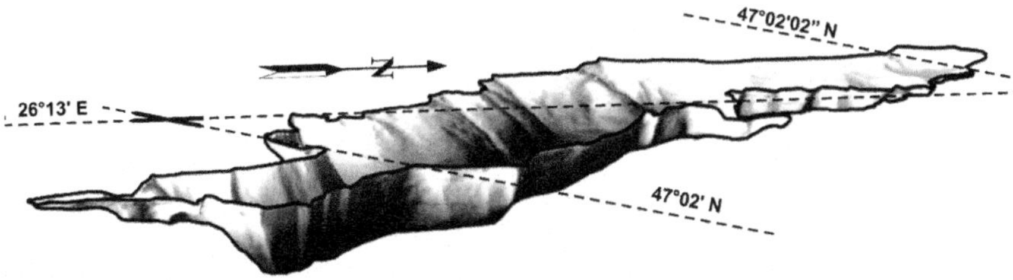

Fig. 5.8 3D model of the Cuejdel lacustrine basin generated in ArcScene software based on bathymetric survey in 2011

The perimeter of the lake or the length of the coastline was determined both by direct measurements, and by cartographical surveying based on the cartographic products and the orthophotoplans. Thus, the maximum length of the coastline was 2950 m (in 1991—2.9 km), a value which was now reduced by approx. −400 m (in 2013—2.5 km) because of negative variations of the levels. Regarding the development of the shoreline or the sinuosity coefficient, calculated based on the ratio between the perimeter and the square root of the circle of area equal to that of the lake at the time of measurement, the results show a more pronounced festoon of the initial banks (in 1991—4.13) than the current (in 2011—2.11; 2013—3.85). It should be noted that on the lacustrine surface there are not islands or lacustrine alluvial accumulations in the form of banks that emerge to surface, which does not influence this coefficient. For this reason, banks uniformity over time is a consequence of water dynamics which shaped the slopes of the lacustrine basin, both by the wave action along the coastline, and through the general direction of currents inside the water body.

The oscillation of the water level variation directly affects morpho-bathymetric parameters characterizing a lake. In the case of Cuejdel Lake at the end of 1991, the relative altitude of the water mirror was 665 m, given that within the topographical plans was given a minimum altitude of the topographic surface of 645–646 m, equivalent to the maximum depth. In spring 1992, the level fell abruptly by approx. 0.5–0.6 m, due to repeated floods that changed the mouth of the lake. After that time, the lacustrine accumulation entered into a state of relative hydromorphological equilibrium, but with a slight downward trend. In 2011, the water surface elevation was 661 m, and one year later the values determined using an advanced GPS showed an altitude drop of approx. 0.5–1 m (Fig. 5.9).

Level oscillations are best highlighted by the maximum depth depreciation, which currently is reflected approx. 50 m behind the natural dam in the sector where the old valley initially narrowed form of carved gorges. Within the lacustrine basin reconstruction based on topographic plans and scans using ground penetrating radar (GPR) technology (Malå Ramac X3 M and 100 MHz antenna) maximum depth reached 18.8–19 m. In the year 1992, the bathymetric sketch highlighted a maximum depth of only 16.3 m. In 2011, this parameter indicated on the interface of the echo-

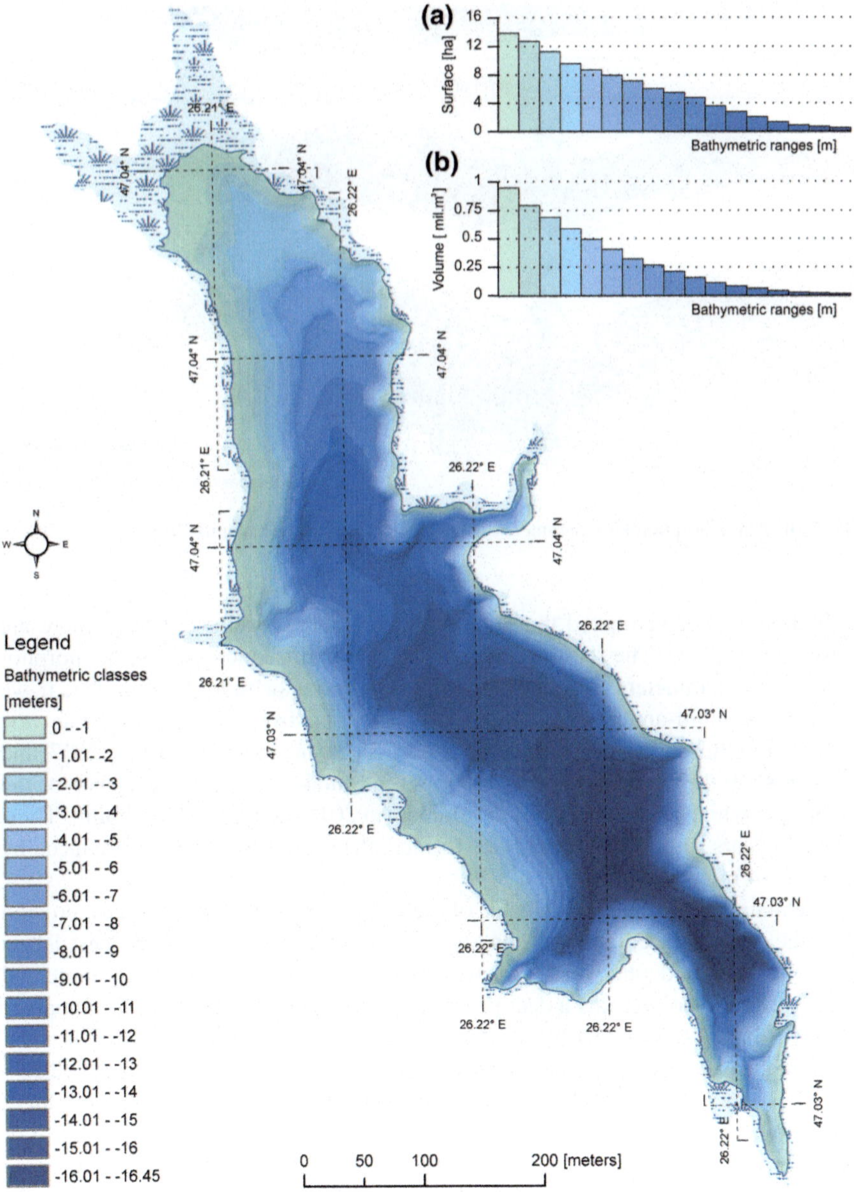

Fig. 5.9 Bathymetric map of the Cuejdel lacustrine basin in 2013; **a** Characteristic isobaths and equivalent value [m^2] of the area; **b** Cumulative volume [m^3] for bathymetric intervals of −1 m depth (Mihu-Pintilie et al. 2012, 2014a, b, c, d, 2016)

sounder used by our research team a value of 16.45 m, and in winter of 2013, 14.9 m (Mihu-Pintilie et al. 2014a).

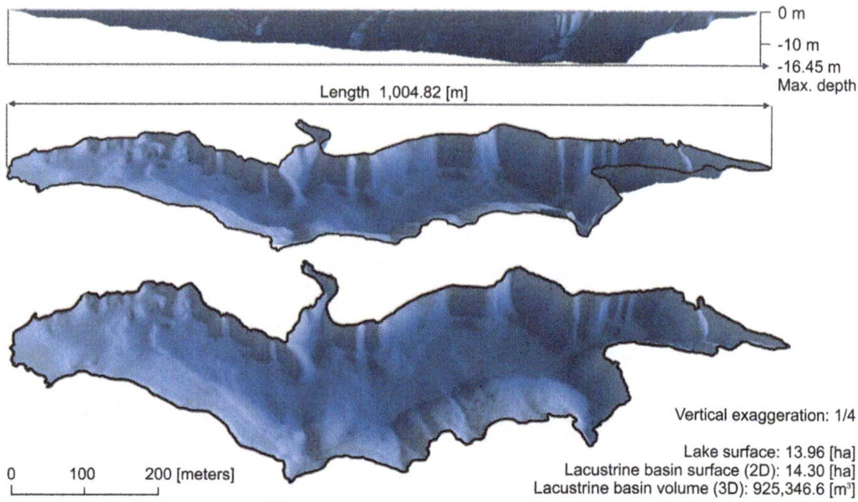

Fig. 5.10 Main morphometric parameter of the Cuejdel lacustrine basin in 2013

It should be remembered that in addition to annual level oscillations there are seasonal variations. The seasonal oscillations may induce increases of the morpho-bathymetric parameters, especially in terms of floods or rainfall. However, if we consider the multi-annual averages, they indicate negative variations of the aquatic surface. For this reason, measurements performed in winter (in 1992, 2013 and 2014) restore the minimum due to the low fluid intake, and to the presence of the ice bridge, while work carried out in summer (in 2011) restore the maximum. Thus, in the 25 years since the Cuejdel Lake formed, the maximum depth was reduced by approx. −4 m.

The main cause of the loss in value of the maximum depth depends on the erosion from the dam surface, but isalso due to accelerated clogging of the deepest lake's area. The average depth, which is determined by the ratio between the volume and water surface, also presents a decrease trend, falling at the same time with approx. 1.7 m (in 1981—2.0 m; in 1991—7.54 m; in 2011—5.6 m; in 2013—5.8 m).

From the category of morpho-bathymetric parameters, the actual size of the lacustrine depression floor, which expresses the total submerged area covered by direct mass of water at a time, can be expressed only through the achievement of DEM of the lacustrine basin. The 3D reconstruction was performed in the ArcGIS software (command Topo to Raster) using the altitudinal levels from the topographic and forest plans made out for the years 1981 and 1991, the bathymetric levels from 2011 and the GPR profiles from 2013 (12 transverse profiles and one longitudinal profile). Thus, the largest surface occupied by water was of 165,646.9 m^2, now this parameter decreasing from 143,188.8 m^2 (in 2011) to 140,942.1 m^2 (in 2013), with a total emerged surface in the examination time period of approx. 24,705 m^2 (Fig. 5.10).

The volume of water in the lake was reduced with the increase of the liquid output, caused by the enhancement of the overland flow on the surface of the dam, to which were added the water losses caused in time by the deepening of the drainage channel or by the infiltrations through the mass of the dam. The estimation of the water volume to the initial lake was carried out on the basis of the bathymetric information extracted from the forest plans (ed. 1981, 1:10,000), summing the partial volumes from between two isohypses, respectively two isobaths. For calculation the Simpson rule has been used (Eq. 5.1).

Equation (5.1), Simpson rule for volume calculation:

$$\mathrm{V} = \mathrm{h}/3 * (S_0 + 4S_1 + 2S_2 + 4S_3 + 2S_4 + \ldots + S_n) \ \left[m^3\right] \tag{5.1}$$

where: h—the difference in depth between two bathymetric curves, S_1–S_2…S_n—surfaces of bathymetric curves (Minea and Romanescu 2007).

Thus, in 1981 the volume of reconstituted water volume of Cuejdel Lake was approx. 4.5×10^4 m^3, representing approx. 3.6% of the accumulated water after the events of the summer of 1991.

Counting the volumes variations from the current lake emergence (1991) to date was based on the analysis of the numerical model of the lacustrine cuvette in ArcGIS software (command 3D Analyst Tools—Functional Surface—Surface Volume), which offers the possibility of extracting both total and partial volumes for each bathymetric level. In the year 1991 the volume estimated was 122.3×10^4 m^3 ($\simeq$ 1223,001.33 m^3) and was reached when the accumulated water in the newly created lacustrine depression exceeded the deluvial dam. In 2011, direct measurements determined a water volume of approx. 92.53×10^4 m^3 ($\simeq$ 925,346.55 m^3), and two years later, in winter conditions with some of the liquid mass retained in the ice bridge, a volume of 81.6×10^4 m^3 (815,986.01 m^3) (Table 5.1).

The water losses between 1991 and 2011 were approx. 29.77×10^4 m^3 (297,654.77 m^3), to which may be added, but only as seasonal variation, the difference between 2011 and 2013, which was 10.94×10^4 m^3 (109,360.54 m^3). The decline of water level shows a downward trend of values in all morphometric and morpho-bathymetric parameters, caused by surface morphological changes. In the case of the volume losses it can be added also the clogging rate of the lacustrine basin, which increases over time inversely proportional to the water mass (Fig. 5.11).

The hypsometrical integral that describes in one value the shape of lacustrine basin based on the ratio between water depth (y = h/H, where: h—depth interval between two isobaths, H—max. depth of lake) and the surface of the isobaths (x = a/A, where: a—isobaths area, A—lake surface), also indicates a clear tendency of morphological balancing (Strahler 1952; Romanescu et al. 2013).

Thus, at the time of water accumulation over an irregular relief the hypsometric integral presents old thresholds of abrasion terraces carved by initial accumulation. After 20 years, the lacustrine bottom changed its shape due to accretion processes. Considering that the measurements were carried out in winter, when the ice bridge controls the turbidity of the water body, during 2013, the curve of hypsometric integral indicated a more balanced slope (Fig. 5.12).

Table 5.1 Variation of morphometric and morpho-bathymetric parameters of the Cuejdel lacustrine basin between 1981 and 2013 (Mihu-Pintilie et al. 2014a)

Lacustrine basin parameter	Unit of measurement	Year			
		1981	1991	2011	2013
Altitude of lake surface	[m]	648	665	661	660
Lake surface	[$10^4 m^2$]	2.29	16.22	13.95	13.88
Lacustrine basin surface	[$10^4 m^2$]	2.42	16.56	14.31	14.09
Length	[km]	0.451	1.170	1.004	0.987
Avg. width	[m]	50.2	138.6	138.8	140.5
Min. width	[m]	5.0	24.1	19.25	18.2
Max. width	[m]	125.1	341.2	282.6	270.2
Major axis	[m]	410.3	1154	973.7	956.9
Small axis	[m]	105.7	358.4	299.5	285.3
Form coefficient	–	0.25	0.31	0.30	0.29
Perimeter length	[km]	0.957	2.950	2.801	2.544
Sinuosity coefficient	–	3.56	4.13	2.11	3.85
Avg. depth	[m]	5.5	18.8	16.45	15.9
Max. depth	[m]	2.0	7.54	5.6	5.8
Volume	[$10^4 m^3$]	4.5	122.3	92.53	81.6

The hypsographic curve or the isobaths surface/depth (%), respectively cumulative volume/depth (%) ratio is a graphical construction estimating the major changes in the shape of the lake basin in time (Straškraba 1980; Kalff 2003). In 1991, the lake Cuejdel had a surface/linear depth ratio [f(0.5)–f(1.5)], equivalent to proportioned form, and a cumulative volume/convex depth ratio [f(−0.5)–f(−1.5)], that proves thresholds in the bathymetric range of 10–16 m, certainly induced by the banks of the flooded stream (Fig. 5.13).

Currently, the most important changes occurred only to isobaths surfaces, following the constant losses of water, the slightly convex shape of hypsographic curve [f(0.5)–f(−0.5)] showing a reduction of lake bottom relative to its total area. Regarding the cumulative volume/depth ratio, in the 22 years of the evolution of the lake, the original proportions have not changed significantly (Neumann 1959; Håkanson 1981) (Fig. 5.14).

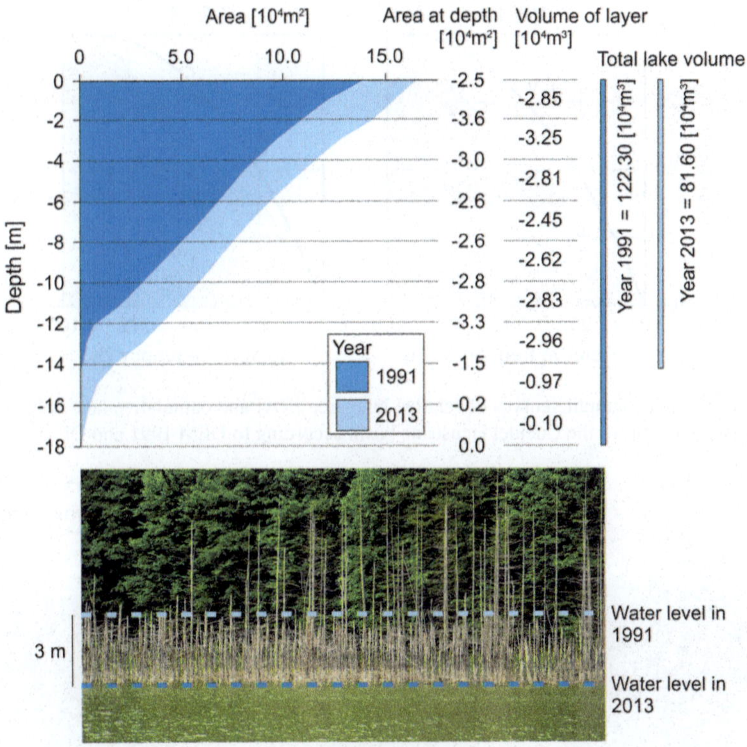

Fig. 5.11 The hypsographic curve of Cuejdel lake indicating the changes in terms of depth, area and volume between 1991 and 2013. The picture is illustrating water level loss marked on the flooded trunks (Mihu-Pintilie et al. 2016)

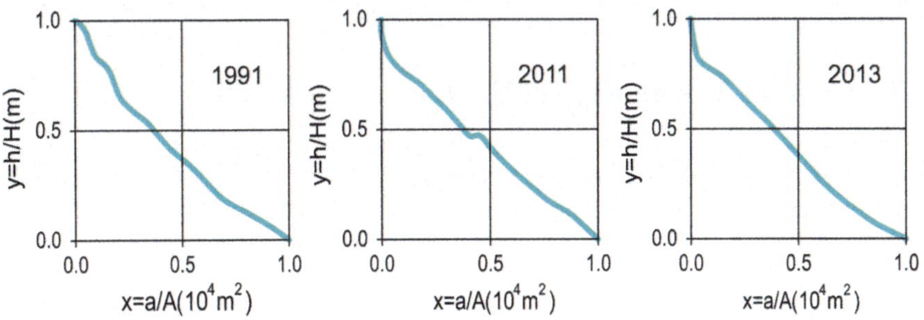

Fig. 5.12 The hypsometric integral of the Cuejdel lacustrine basin in 1991, 2011 and 2013 ($y = h/H$, where: h—depth interval between two isobaths, H—max. depth of lake; $x = a/A$, where: a—isobaths area, A—lake surface) (Mihu-Pintilie et al. 2016)

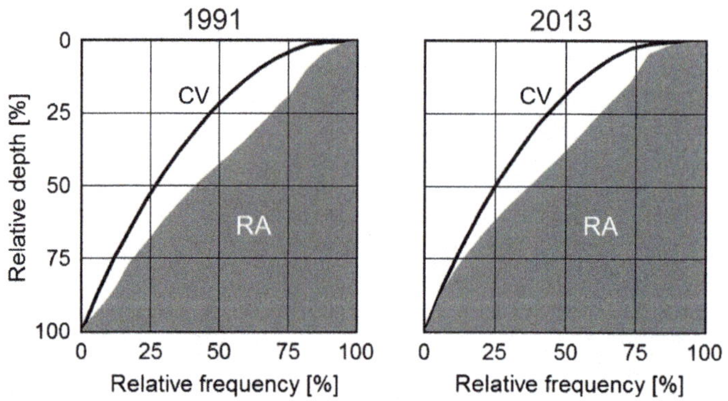

Fig. 5.13 The hypsographic curves of relative lake area [RA] and cumulative lake volume [CV] versus relative depth for the Cuejdel lacustrine basin according to DEM 1991 and DEM 2013

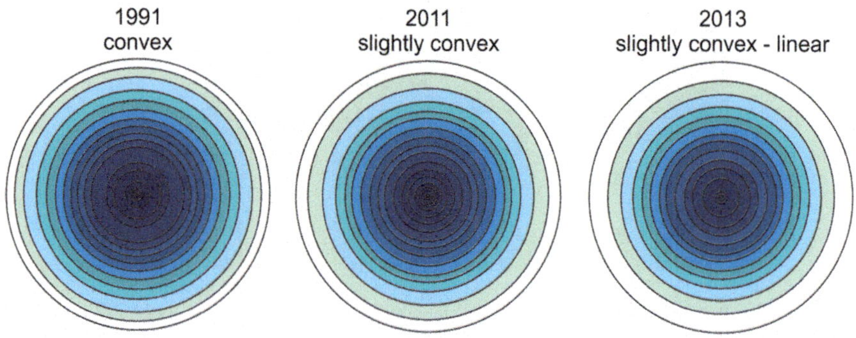

Fig. 5.14 Schematic bathymetric interpretation of the Cuejdel lacustrine basin forms in 1991, 2011 and 2013

5.3 Current Morphology of the Lacustrine Basin

The morphology of the Cuejdel lake basin is closely linked to the formation and development of the flooded valley where it took major geomorphological features, such as: shore and coastal zone, slopes and submerged valleys, lake bottom, meander-shape due to the old watercourse etc. To these were added the alluvium carried in lake by the river system and the materials from both the banks erosion and from the biodegradation of the aquatic vegetation in situ. The totality of the accumulated deposits was decanted on the bottom of the aquatic surface where they were modeled in time by the water dynamics (waves, currents, floods) (Mihu-Pintilie et al. 2012) (Fig. 5.15).

Fig. 5.15 Inflow and outflow of Cuejdel Lake: **a** Glodu brook; **b** Cuejdel brook; **c** Outflowing branch of lake over the dam surface (Mihu-Pintilie et al. 2016)

To describe in more detail how the current macro-relief submerged of the current lake cuvette the map from 2011 was used. The map was created based on bathymetric data measured by the echo-sounder, from which it resulted by interpolation in GIS, a DEM with a resolution of 0.5 m/pixel. The main morphological elements were: the slopes of the lake—where the shore (the supralittoral zone), the littoral zone and the littoral bank—the lacustrine bottom and the deluvial dam are noticeable.

The shore or the supralittoral zone correspond mainly with the emerged sector due to negative variations of levels and have different geomorphological characteristics depending on the structure of the original valley. The right bank, between Rotund creek and the Glodu creek, has a slope of between 5° and 20°, a width of 10–15 m up to the forest and its fragmentation is induced by Piciorul Crucii brook and a number of torrents which feed the lake during heavy rainfall. A similar configuration can be found on the opposite bank, in the sector between Cuejdel creek and Făgetu creek, except that here the degree of fragmentation is lower due to the forest road from the median part of the left slope, which slows the development of the erosional landforms.

The shore sector between the right side of Făgetu creek and up close to the ancient waterway of the lake reveals less fragmentation, but more abrupt sides (20°–50°), resulting in a series of gravitational processes (landfalls, rollings etc.) potentiated in the basis by the lacustrine erosion. However, the forest coverage degree to just above the water mirror results in bank failures on restricted sectors. From this point of view, however, the southwest bank stands out, between the current waterway of the lake and Piciorul Rotund creek, where banks have the highest slope (45°–60°) and present scarps induced by the lacustrine abrasion.

The north bank, which corresponds to the upper lacustrine basin has undergone the biggest transformation induced by the progressive withdrawal of the aquatic surface. Currently, between the old coastal terrace and the limit of the aquatic surface a distance of 50–170 m has been created, where the slope does not exceed 5° as a result of the emerged alluvial fan that has evened out the slopes. The relief fragmentation is high due to the streams Glodu and Cuejdel, which had disheveled their beds into the Fan-delta type. The largest wetland area around the lake was installed here, where hygrophilic vegetation grows to the detriment of unadapted trees to aquatic life.

The coastline festoon is given mainly by the mouth of the streams which due to the water accumulation turned into short gulfs. The most important of these is the Făgetu stream's outlet, which, due to its deep riverbed (5–8 m) and its relatively narrow transverse profile (approx. 50–60 m), has become through flooding estuary-shaped. Banks in this sector exceed 30°–35° and the average bed has a slope of 10° pursuing its submerged route to the old confluence with the Cuejdel creek. The mouths of the Piciorul Crucii and Pârâul Rotund creeks are slightly wider, which caused a wider strandline and slightly shorter submerged valleys.

The lake littoral is considered that section festooning the coastline, and its lower limit is given by the depth to which underwater plants are growing. In the case of Cuejdel Lake an upper flood littoral zone is less predominant, as in the case of large lacustrine surfaces because the level oscillations during this period of hydro-morphological balance are small and the emerged sectors were invaded by terrestrial vegetation. Best highlighted is the lower littoral, represented by the alluvial bench built of accumulation materials resulting from the narrowing of the shores. From this point of view, the right bank of the lake develops a lower littoral zone with a width of 20–40 m formed on the less abrupt slope of the old valley. An important role in the alluvial delivery of the littoral bank is played by the submerged forest that prevents eroded materials from settling in the deep lake area. After the 3 m isobath, from where the slope increases from 3°–5° to 10°–15°, the littoral area stops, and the transition to the deep occurs through a micro-terrace that inherits the morphology of the left bank of the Cuejdel stream (Fig. 5.16).

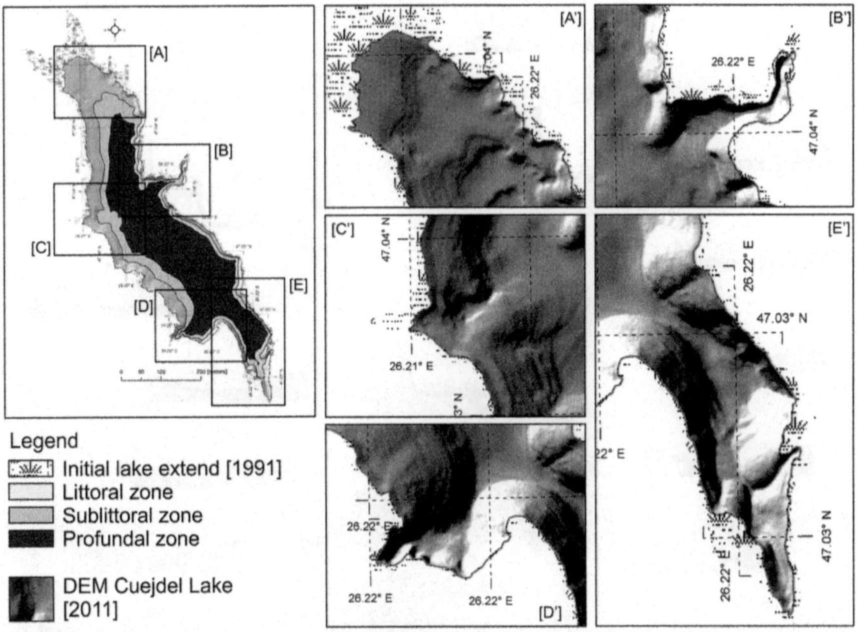

Fig. 5.16 Zoning of the Cuejdel lacustrine environment and current morphology: [A-A′] Upper lacustrine basin and mouth stream of Glodu and Cuejdel brooks; [B-B′] Mouth stream of Fagetu brook on the left bank of lake; [C-C′] Mouth stream of Piciorul Crucii brook on the right bank of lake; [D-D′] Mouth stream of Rotund brook and the confluence with Cuejdel brook; [E-E′] Lower lacustrine basin and area with maximum depth of Cuejdel lake (Mihu-Pintilie et al. 2016)

The lower littoral area from the left side and that between the Rotund creek and the lake waterway is short because of the abrupt submerged slopes, with declivities of 35°–50°, in some sectors even exceeding 60°. Also, throughout this segment the transition from the shore limit to the lake bottom occurs suddenly, the depths reaching 14–15 m over a distance of only 5–7 m from the shore. Within the sector with the greatest depths behind the dam (>14 m), this morphology is evidenced by the slopes of the old gorge sector carved by the Cuejedel creek until the basis of Piciorul Rotund. Only in the abandoned waterway a small productive portion has developed due to the formation of an alluvial bench powered by the overland flow discharging alluvial deposits before leaving the lake (Fig. 5.17).

The area with the largest littoral zone is found in the upper lake basin, where the submerged alluvial fan is continuously fed by Glodu and Cuejdel creeks, and aquatic vegetation is supported by tree trunks caught in the water. Until the 3 m isobath, from where the depth begins to grow progressively, the bottom of the lacustrine depression is productive because of the low slope (1°–5°) and of the leveling of the initial relief microforms through clogging. After this sector, there are a number of thresholds corresponding with alluvial lobes foreheads that follow the direction of the currents

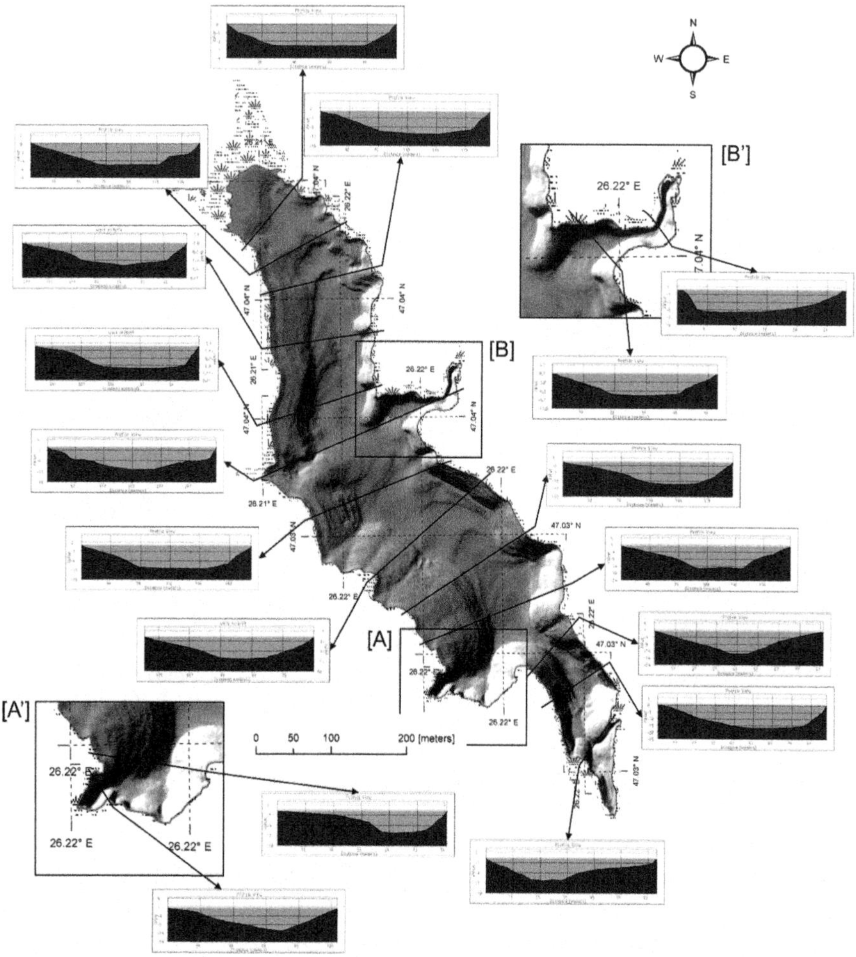

Fig. 5.17 Cross sections through the Cuejdel lacustrine basin; [A-A′] View over mouth stream of Paraul Rotund; [B-B′] View over mouth stream of Fagetu (Mihu-Pintilie et al. 2016)

over the alluvial micro-terraces, most corresponding to those formed on the right side.

The deep area of the lake is better underlined through the longitudinal profile, where the asymmetry of the slopes is outlined showing the emergence of the bottom of the aquatic surface. From the input zone of accumulation to the dam basis correspondent with the maximum depth of 16.45 m measured in 2011, the slope has a length of approx. 855 m and is dominated by accumulation processes. The general slope of 19° is accentuated by alluvial micro-terraces appearing as small thresholds that connect the littoral zone to the denuded bottom (Fig. 5.18).

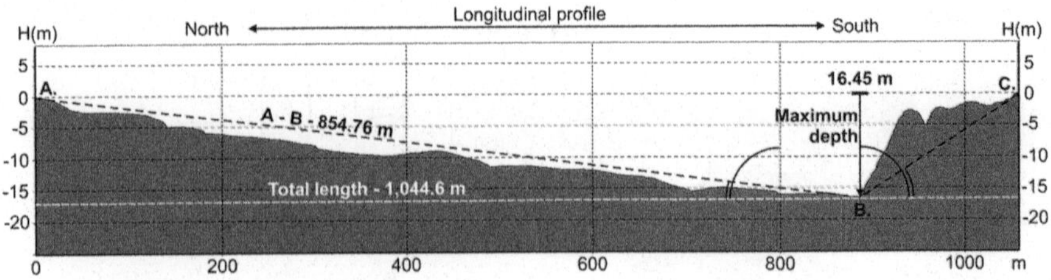

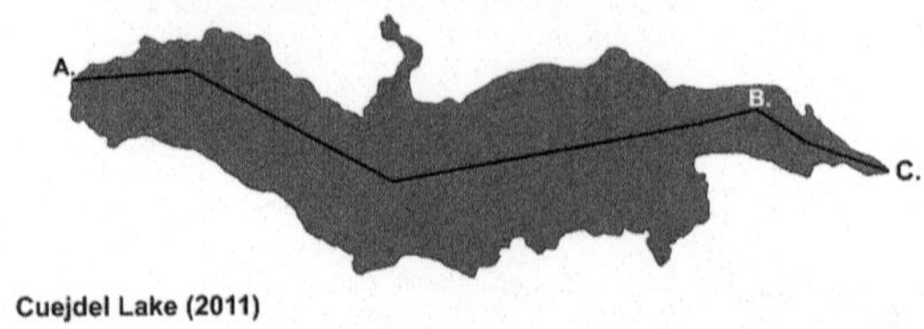

Segment	A - B	B - C
slope α (degree)	19	34
slope (%) = tgα° 100	34.43	67.45
slope (‰) = (Hmax - Hmin)/L	0.019	0.086
depth - Hmin (m)	0	0
depth - Hmax (m)	16.45	16.45
length - L (m)	854.76	189.84

Fig. 5.18 Longitudinal profile in the Cuejdel lacustrine basin (Mihu-Pintilie et al. 2016)

In contrast, the slope located between the bathymetric maximum and the waterway is short (approx. 190 m) and abrupt (34°), with a fragmented morphology due to the chaotic disposal of deluvial material that blocked the valley. On the surface of this slope, the morphology is continually changing as a result of alluvial accumulations at the mouth of the lake, leading to the lake bottom ascension within the central sector and to the left bank. To the right bank the flow channel led to deepening the lake bottom profile, accentuating the asymmetry of the slopes in this sector.

According to the histogram of the slopes map, the dominant declivity class is 5°–10° (25.67%), while the less represented exceeds >50° (0.33%). With a high percentage are slopes characterized by the interval 1°–3° (18.95%), and 3°–5° and 15°–25° classes divide approx. equally 28% of the lacustrine depression.

Low declivity (0°–1°) characterizes the bottom of the water surface uniformed by the accumulation processes (7.5%), and the submerged bench foreheads are highlighted by the slope class of 10°–15° (11.13%). The connection between the lower littoral zone and the bottom of the lake, and the submerged sector of the natural dam is made by abrupt slopes (25°–60°) totaling 8.9% (Figs. 5.19 and 5.20).

In Romania, there are around 3450 lakes of which 2300 are natural and 1150 are man-made reservoirs, with a total surface of about 2620 km^2 (Gâștescu 1971; Romanescu et al. 2013). Analysing the morpho-bathymetric parameters of the 176 main lakes from the Carpathians Mts., of which 112 with natural origin (82 glacial lakes, 25 natural dam lakes, two cryo-nival lakes and one volcanic crater lake) and 65 man-made reservoirs (except Izvorul Muntelui Reservoir which due to its large size it is out of statistic range), we have identified a number of related data between elevation (m), max. depth (m) and water surface (m) (Fig. 5.21).

The maximum depth of the lakes in the mountain areas ranges between −1 m and −180 m. The deepest (more than −40 m) are specific to the man-made reservoirs, which were built in the Communist Era in the interior valleys of the Carpathians, between 500–1500 m altitudinal range. Natural lakes do not exceed the depth of

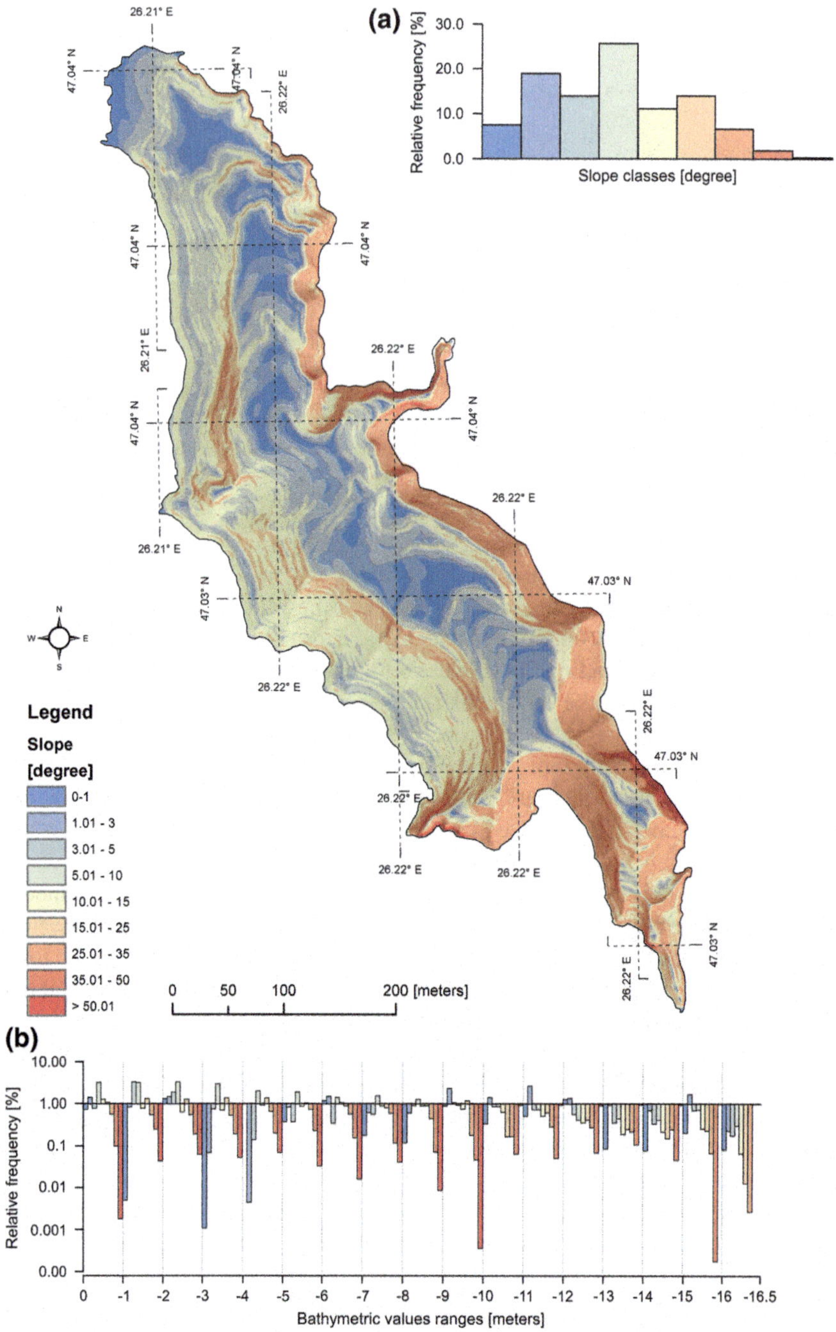

Fig. 5.19 Slope map of the Cuejdel lacustrine basin; **a** The average distribution [%] of slope classes; **b** The average distribution [%] of slope classes for each 1 m depth interval

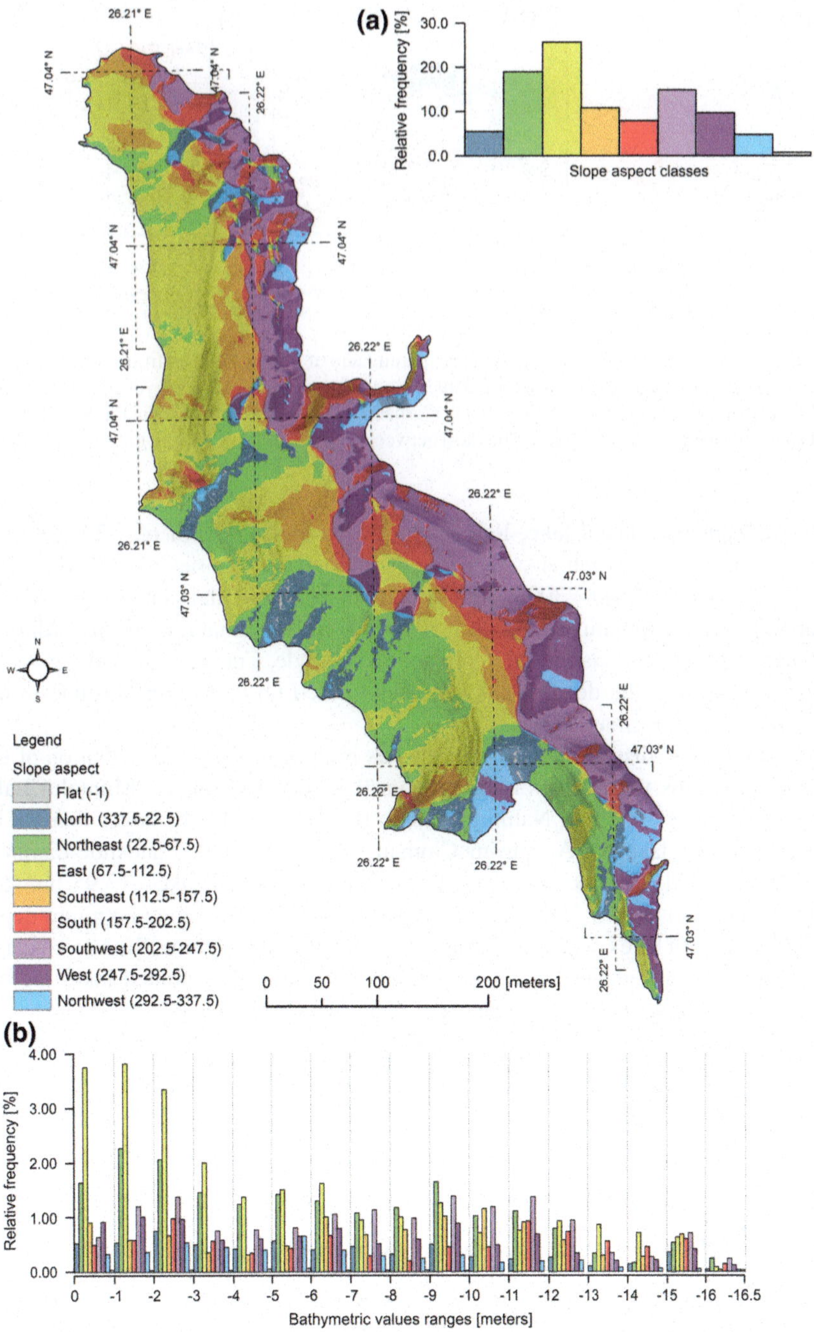

Fig. 5.20 Slope aspect map of the Cuejdel lacustrine basin; **a** Average distribution [%] of banks aspect clasess; **b** Distribution [%] of banks aspect for each depth interval

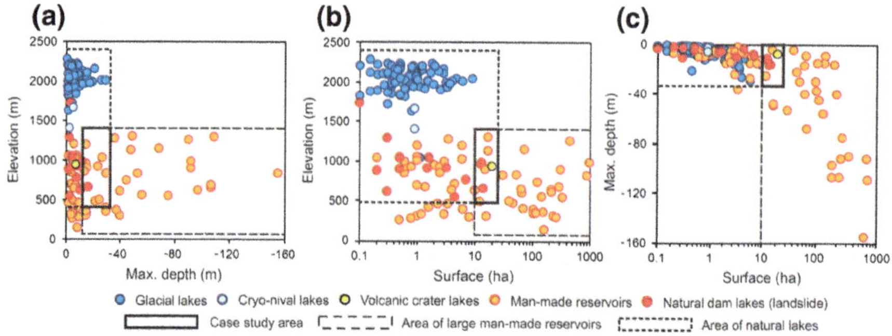

Fig. 5.21 The morphometric and bathymetric parameters of 176 main lakes from which 112 with natural origin (82 glacial lakes, 25 natural dam lakes, two cryo-nival lakes, one volcanic crater lake) and 65 man-made reservoirs from Romanian Carpathians. The distribution of maximum depths **a** and lakes surface **b** on the altitude; **c** The ratio between the maximum depth and the surface of lakes

−30 m (Zănoaga glacial lake, Retezat Mts. with a max. depth of −29 m); their distribution on the altitudinal steps is divided into two categories: glacial and cryo-nival lakes (within 1500–2300 m altitudinal range), natural dam lakes and volcanic crater lake from Ciomatu Mts. (within 500–1500 m altitudinal range). According to the distribution of the maximum depths on the altitude, Cuejdel and Red landslide-dammed lakes are the deepest natural lakes formed up to 1500 m altitude in the mountain area of Romania (Fig. 5.21a).

The altitudinal distribution of water bodies is similar with the distribution of the maximum depths with values ranging from 0.5–1000 ha (Izvorul Muntelui Lake with a surface of 3000 ha). Natural lakes do not exceed surfaces larger than 20 ha (volcanic crater lake Sf. Ana from Ciomatu Mts.—19.5 ha). Landslide-dammed Cuejdel and Red lakes occupy the next positions (Cuejdel Lake—13.95 ha, Red Lake—12.01 ha) (Fig. 5.21b). According to the ratio between the maximum depth and the surface of lakes, which is also an estimation of water volume, Sf. Ana volcanic crater lake, Bucura glacial lake, Cuejdel and Red landslide-dammed lakes are the most representative natural lakes in Carpathian Mts (Fig. 5.21c).

References

Gâştescu P (1971) Lacurile din România–limnologie regională. Romanian Academy Publishing House, Bucharest

Grigore M, Ielenicz M (1972) Cartografierea porniturilor de teren. Bull Soc Sci Geogra 1(70):236–242

Håkanson L (1981) A Manual of Lake Morphometry. Springer, Berlin

Ichim I (1972) Văile elementare din Munții Stânișoarei și raportul lor cu structura și litologia. Stud Cerc Geol Geofiz Geogr, S Geogr 19(2):175–182

Ichim I (1973) Rôle de la structure et de la lithologie dans la morphologie des vallées des Monts Stânişoarei. Stud Geomorph Carp–Balc 7 (Bucharest)

Ichim I, Rădoane N, Rădoane M (1996) Procese geomorfologice cu intervale de recurenţă mare în arealul munţilor flişului. Exemplificări din Judeţul Neamţ (Geomorphological process with high frequencies in flysch mountains area. Examples from Neamţ County). In: Studii şi cercetări—extras 8, Muzeul de Ştiinţe Naturale Publishing, Piatra-Neamţ, p 15

Kalff J (2003) Limnology-Inland Water Ecosystems. Prentice-Hall, Upper Saddle River, New Jersey

Mihu–Pintilie A, Romanescu G (2011) Morphometric and morphological suitability of the relief from the Crucii (Cuejdel) Lake basin (Stânişoarei Mountains). In: Pandi G, Moldovan F (eds) Proceedings of the air and water components of the environment, vol 1. Babeș–Bolyai University, Cluj, 18–19 March 2011, pp 305–313

Mihu-Pintilie A, Romanescu G, Stoleriu C (2012) Morpho–bathymetric parameters of recess Crucii (Cuejdel) Lake (Stânişoarei Mountains). In: Pandi G, Moldovan F (eds) Proceedings of the air and water components of the environment, vol 1. Babeș–Bolyai University, Cluj, 23–24 March 2012, pp 445–452

Mihu-Pintilie A, Romanescu G, Stoleriu CC, Nicu IC, Asăndulesei A, Schmaltz E (2014a) Natural dam lakes from Cuejdiu watershed (Stânișoarei Mountains)—Non–invasive methods used for bathymetric maps. In: Gâştescu P, Marszelewski W, Breţcan P (eds) Proceedings conference of the water resources and wetlands, vol 1. Tulcea, 11–13 September 2014, pp 130–137

Mihu-Pintilie A, Romanescu G, Stoleriu C (2014b) The seasonal changes of the temperature, pH and dissolved oxygen in the Cuejdel lake. Romania. Carpat J Earth Enviro Sci 9(2):113–123

Mihu-Pintilie A, Paiu M, Breabăn IG, Romanescu G (2014c) Status of water quality in Cuejdi hydrographic basin from Eastern Carpathian, Romania. In: SGEM–Proceedings of the 14th international multidisciplinary scientific geoconferences—hydrology and water resources, vol 14, issue 1. 18–24 June, Albena, pp 639–646

Mihu-Pintilie A, Romanescu G, Stoleriu C, Stoleriu O (2014d) Ecological features and conservation proposal for the largest natural dam lake in the Romanian Carpathians-Cuejdel lake. Int J Conserv Sci 5(2):243–252

Mihu-Pintilie A, Asăndulesei A, Nicu IC, Stoleriu CC, Romanescu G (2016) Using GPR for assessing the volume of sediments from the largest natural dam lake of the Eastern Carpathians: Cuejdel Lake, Romania. Environmental Earth Sciences 75:710. https://doi.org/10.1007/s12665-016-5537-1

Minea I, Romanescu G (2007) Hidrologia mediilor continentale–Aplicaţii practice. Demiurg Publishing House, Iaşi

Neumann J (1959) Maximum Depth and Average Depth of Lakes. Can Fish Res Bd 16:923–927

Rădoane N (2002) Un nou lac de baraj natural în bazinul Bistriţei Moldoveneşti-Lacul Cuejdel. Stud Cerc Geograph 49:76–82

Romanescu G, Stoleriu C, Enea A (2013) Limnology of the Red Lake, Romania. An interdisciplinary study. Springer, Dordrecht, Heidelberg, New York, London

Rusu C, Mărgărint MC, Rusu E, Boamfă I (2002), Noi arii de interes ecologic din Moldova: Parcul Forestier Vânători—Neamţ şi Lacul Crucii (bazinul hidrografic al Cuejdelului). Revista Terra, 1–2 (Bucharest)

Stoleriu C, Stoleriu O, Mihu-Pintilie A (2014) Scientific and tourist value of natural dam lakes in the Carpathian Mountains (Romania). Case study: Red, Cuejdel and Iezerul Sadovei Lakes. In: SGEM–Proceedings of the 14th International Multidisciplinary Scientific Geoconferences–Ecology and Environmental Protection, vol 14, issue 2. 18–24 June, Albena, pp 625–632

Strahler AM (1952) Hypsometric (area-altitude) analysis of erosional topography. Bull Geol Soc America 63:1117–1142

Straškraba M (1980) The effect of physical variables of freshwater production: analyses based on models. In: Le C, Lowe-McConnell RH (eds) The functioning of freshwater ecosystems, vol 1. Cambridge University Press, Cambridge pp 13–84

Surdeanu V (1986) Landslides and their role in reservoir silting. Z Geomorph NF Suppl–Bd 58:165–171 (Berlin–Stuttgart)

Chapter 6
Nature of Lacustrine Sediments and Clogging Rate

Abstract In this chapter sediment discharge and the clogging rate are analysed in order to estimate the lifespan of Cuejdel Lake. The silting processes were identified through indirect calculation methods of soil erosion (USLE) and through geo-physical investigations (direct methods). Based on the indirect methodology of alluvial budget, the result indicates for the catchment area an alluvial production of 3288.5 m^3/year (3.02 t/ha/year), in 23 years reaching ca. 75,635.93 m^3. From the direct investigation methods, we mention the ground-penetrating radar (GPR) technology, a geo-physical tool accepted long time ago in scientific practice, but unfortunately, shortly used in limnological studies. The bathymetric measurements were concretized by the reconstruction of flooded valley and the achievement of the map of thickness sediments accumulated in the lake basin. In the present time, the water volume was reduced with 297,654.77 m^3, from which 103,988.87 m^3 were replaced by sediment deposits. The sediment thickness is between 0.1 and 3.6 m. The thickness class 0.1–1.0 m corresponds to 8.552 ha (60.7%) from the lake bottom, 1.0–2.0 m corresponds to 4.676 ha (33.32%), and the thickness class >2.01 m covers the rest of 8326.5 m^2 (5.93%). The clogging thickness rate is between 1 and 16 cm/year which means a medium value of 4521.26 m^3/year. This value, related to the catchment basin (8.94 km^2) and to an effluence rate of 33%, indicates a medium production of sediments—4.33 t/ha/year.

6.1 The Catchment Lake Basin and the Sediment Budget Evaluation

The alluvial deposits from Cuejdel Lake were formed through two major sedimentary processes. The first process occurs by the accumulation in situ due to the volume of water dynamics and pressure exercised on the submerged slopes and on the banks. In areas where the slope exceeds 25°–30°, mud lobes are forming and slid up on the surface of lacustrine micro-terraces towards the lake bottom. Alluviation also occurs due to water level decrease and the sediment intake brought by permanent tributaries (Glodu, Cuejdel, Făgetu, Piciorul Crucii and Rotund creeks) or by torrents. In this way, the fine deposits are returned to the lacustrine basin. The second and most

A. Mihu-Pintilie, *Natural Dam Lake Cuejdel in the Stânişoarei Mountains, Eastern Carpathians*, https://doi.org/10.1007/978-3-319-77213-4_6

Fig. 6.1 Water turbidity of Cuejdel Lake after a flash flood (12 September 2012). **a** Inflow of Cuejdel and Glodu streams; **b** Fan-delta formed in the upper sector of the lake; **c** Outflow of the lake—natural dam area

important source of sediments comes from the catchment area and is produced by the sediment transport within the lake through the network of drainage or by areolar drain (Fig. 6.1).

In hypsometric terms, the catchment basin of Cuejdel Lake (surface—8.94 km^2) is characterized by elevations between 1117 m (Muncelu peak) and 661 m (lake level), with the highest participation of altitudes up to 850 m (>70%) (Fig. 6.2a). The main peaks are imposed by the lithological structure made of alternating sedimentary rocks specific to Carpathian external flysch area (Oligocene in west—sandy complex, shale and conglomerates; Cretaceous-Paleocene in the east—shaly flysch, limy flysch, limy-shaly or sandy flysch, clays). The valleys are grafted on more permissive lithological deposits specific to the interval Paleocene—late Eocene, sometimes rocks occurring in minor beds due to erosion. The entire complex consists of an

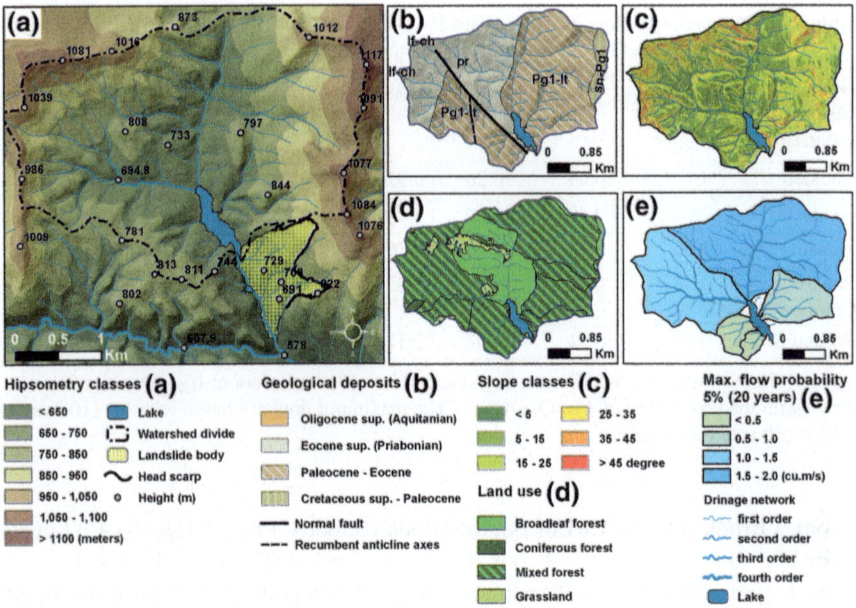

Fig. 6.2 Thematic maps of the Cuejdel catchment basin: **a** Hypsometry; **b** Geology; **c** Slope; **d** Land use; **e** Maximum flow discharge with 5% insurance (20 years) (Mihu-Pintilie et al. 2016)

arrhythmic disposal of layers of shaly, shaly-sandy or sandy flysch, with deposits of red, green and gray-green clays, marly sandstones and marls (Mihu-Pintilie and Romanescu 2011) (Fig. 6.2b).

The soil layer from the top has a reduced thickness (0.5–1 m), clayey and sandy texture and a moderate quantity of humus (3.9–13.5%). These characteristics are specific to cambisoils (>85%) and argillic soils (12%) classes and in a low proportion to undeveloped soils. Considering the litho-pedological context, the slopes between 5° and 61° accelerate erosional processes and increase suspended load of the streams during cloud bursts (Mihu-Pintilie et al. 2014a) (Fig. 6.2c).

96.2% from the catchment lake basin is covered by forests and 3.8% by alpine and sub-alpine meadows. The watercourses have a total length of 4.96 km and have developed a potential drainage network of the 4th order (according to the Horton-Strahler classification) (Fig. 6.2d). The main reason why the lake dam has not been affected by erosion until the present time is due to the afforestation that ceased the development of hydrographic network. The probability of maximum discharge was determined based on the drainage coefficients, with an insurance of 5% ($Q_{5\%}$) (Diaconu 1992). Thus, within the catchment basin, in the last 20 years it was registered at least one event of flash flood with a maximum discharge between 0.5–2.0 m^3/s considering that downstream the lake, at Cuejdiu gauge station, the average discharge of Cuejdi river is only 0.54 m^3/s. The level of alluvial charge in high waters can also be discerned from the hydronymy of the watercourses that flow into the lake: the Cuejdel

Table 6.1 The maximum flow discharge with 1% (100 years) and 5% (20 years) insurance in the catchment basin of Cuejdel Lake calculated based on rational method (Diaconu 1992)

Drainage basin of lake tributary	Surface (Km^2)	α_{Global}	T_c (s)	$I_{1\%}$ (mm/min)	$Q_{1\%}$ (m^3/s)	$Q_{5\%}$ (m^3/s)
Cuejdel	4.2	0.49	70.26	0.92	3.21	1.81
Glodu	2.6	0.491	49.70	1.13	2.42	1.13
Fagetu	1.06	0.51	41.73	1.26	1.13	0.64
Piciorul Crucii	0.38	0.49	19.93	1.93	0.6	0.33
Rotund Brook	0.18	0.51	14.52	2.31	0.35	0.2

α_{Global}—Global water flow coefficient; T_c—Time of runoff concentration; $I_{1\%}$—Average intensity of rain with the probability of 1%; $Q_{1\%}/Q_{5\%}$—The maximum flow discharge with 1% (100 years) and 5% (20 years) insurance

('stony') Brook (Q_5%—1.81 m^3/s), the Glodu ('muddy') Brook (Q_5%—1.37 m^3/s) or the Făgetu ('beech forest') Brook (Q_5%—0.64 m^3/s) (Fig. 6.2e; Table 6.1).

As regards the transfer of sediments in the hydrographical basin, respectively from slopes, in the beds of streams and then into the lake, and the alluvial beddings from the coluvio-proluvio-alluvial deposits, they form the alluvium system of the analysed area (Ichim 1986). In practice, the term is explained by identifying the alluvium sources generated by processes within the catchment area, and their flows. When flows are quantified the sediment budget is obtained.

The sediment budget for the Cuejdel Lake basin is problematic because direct measurements could not be made on the evolution of geomorphological processes, on liquid and solid flows, and on beddings. However, estimation of the sediment budget of the lake basin has been tried starting from the data and results of the geomorphological team from Stejarul Station, who worked in the flysch area of Eastern Carpathians. Their results can be extrapolated for our area of study, but validation will be made in the next section based on the clogged sediment volumes using the GPR method.

The first step in estimating the sediment budget has been identifying the sediment sources. As processes generating alluvium within the lake basin, the following sources can be identified: surface erosion, creep, bank erosion along the drainage network, and erosion due to forest roads. The sediment flow was initially estimated by the model proposed by Vanmaercke et al. (2014) (Eq. 6.1).

Equation (6.1), The sediment flow model (Vanmaercke et al. 2014):

$$E = 72.1 * e^{0.68*PGA} * L^{0.93} \left(n = 38;\ r^2 = 0.8\right) \tag{6.1}$$

where: E—specific erosion ($t/km^2/year$); PGA—seismic activity (maximum peak ground acceleration for 100 years in (ms^2); for Cuejdel lake basin—2.2); L—lithology (lithologic score proposed by Syvitski and Milliman (2007); for Cuejdel lake basin—1, that means rocks specific to flysch).

Table 6.2 The rainfall-runoff-erosion [PP-q-E] regression equations applied for spatial modeling of runoff that causes surface erosion in various scenarios of vegetation cover within some experimental plots from Stânişoarei Mountains (Arghiriade 1977; Rădoane 2002)

Land use	Relation PP-q-E	Experimental area	References
1. The average amount of precipitation PP (mm) in relation with altitude H (m)			
–	$PP = 376.2 + 0.2649 * H - 0.0000647 * H^2$	Central Group of Eastern Carpathians	Romanian Contract Research No. 65/1985
2. Flow q (mm) in relation with rainfall PP (mm)			
Land without vegetation	$q = 0.4354 * PP - 0.0091$	Buhalniţa hydrographic basin	Arghiriade (1977)
Grassland/pasture	$q = -2.12 + 0.15 * PP$	Buba hydrographic basin	Rădoane (2002)
Beech forest	$0.036 * PP + 0.0104$	Buhalniţa hydrographic basin	Arghiriade (1977)
Mixed forest	$q = 0.0762 * PP - 1.1176$	Valea lui Bogdan Basin	
3. Erosion E (m^3/ha) in relation with runoff coefficient ($Cq = q/PP$)			
Land without vegetation	$E = 4931 * Cq^{2-4}355.7 * Cq + 952.49$	Buhalniţa hydrographic basin	Arghiriade (1977)
Grassland/pasture	$E = 812.2 * Cq^{2-5}56.98 * Cq + 8.9937$		
Beech forest	$E = 0.4706 * Cq^2 + 18.009 * Cq - 0.1974$		
Mixed forest	$E = 2 * Cq$		

By this method, the basin of Cuejdel Lake will give the specific erosion of 321.84 t/km^2/year, corresponding to a production of sediments of 3.21 t/ha/year.

Also, for estimating the surface erosion there were used two methods: the regression equations rainfall-runoff-erosion equation (PP-q-E) and Universal Soil Loss Equation (USLE). Based on the digital elevation model (DEM), average rainfall distribution during hot season was modeled, because only they generate runoff processes. Thus, taking into account data obtained by Arghiriade (1977), and Rădoane (2002), within some experimental plots from Stânişoarei Mountains a percentage of 24% was extrapolated, which represents rainfall in summer and generates flow with surface erosion. Spatial modeling of runoff that causes surface erosion was estimated in the first stage based on current forest vegetation cover (>96.2%). Also, erosion modeling was projected for various scenarios of vegetation cover using the relations from Table 6.2.

The results of these equations indicate a production of sediments in the catchment basin of Cuejdel Lake of 211,365.5 m^3/year, which means that in 25 years from the lake emergence it has been produced an amount of 4,861,407.7 m^3 sediments. For

Table 6.3 The correction coefficient of erosion (S) calculated for soil types within catchment basin of Cuejdel Lake (Moţoc et al. 1979)

Soil type	Texture	Erosion type	Erosion class	Coefficient (S)
ECti-LUV	U = Silty sand	High	6	1.2
ECti	U = Silty sand	Low	3	0.8
ECti-li	U = Silty sand	Moderate	4	0.9
DCst	T = Clay loam	Low	1	0.6
DCti-li	T = Clay loam	High	5	1.0

Table 6.4 The value of correction coefficient for vegetation effect (C) on soil erosion within the catchment basin of Cuejdel Lake (Moţoc et al. 1979)

Unproductive land	Degraded pasture			Pasture	Forest		Grassland	
	Low	Moderate	High		–	Density < 0.6	Natural	–
1.2	0.2	0.3	0.8	0.001	0.001	0.01	0.001	0.5

a sediment delivery ratio of 33%, according to these values the lake should have reached an amount of 1,604,264.5 m^3, which represents a greater amount than the original volume of the lake (according to the PP-q-E equation).

For modeling the surface erosion according to the Universal Soil Loss Equation (USLE), elaborated within USDA (1978) and adapted by Moțoc et al. (1979) for Romania, the following formula was used (Eq. 6.2).

Equation (6.2), USLE adapted by Moțoc et al. (1979) for Romania:

$$E = K * L^{0.3} * \left(1.36 + 0.97 * I + 0.138^2\right) * S * C * Cs \quad (6.2)$$

where: K—the erosivity coefficient established by Moțoc et al. (1979) for the Eastern Carpathians, namely 0.15; L—the pixel length (m), which was considered the pixel resolution, 28.5 m when the flow direction was 1, 3, 5, 7, and 42 m when the flow direction was 2, 4, 6, 8; I—the pixel slope (%); S—the correction coefficient for soil erodibility (Table 6.3); C—correction coefficient for vegetation effect (Table 6.4); Cs—correction coefficient for the effect of erosion control measurements who was chosen particulary because there is no intervention in this regard.

The value of soil erosion of the catchment basin of the lake obtained by this method was estimated to 6614.6 m^3/year, which means that in 23 years was produced an amount of approx. 152,137.9 m^3 (according to the USLE).

For the creep processes the equation proposed by Ichim et al. (1998) was taken into account as the most indicated for valuating the volumes of sediments produced through this process. That is Eq. (6.3).

Equation (6.3), The equation of creep erosion proposed by Ichim et al. (1998):

$$PASP = \left(G_r * G_v * G_r * L_r\right) / S_b \quad (6.3)$$

where: PASP—specific alluvial production (t/km^2/year); Gr—thickness of slope deposits (m); Gv—volumetric weight of slope deposits (t/m^3); Cr—creep speed (m/year); Lv—length of slopes in direct contact with the bed (m); Sb—the catchment area (km^2).

The creep speed was considered as an average of 3 cm/year after Rădoane (2002), and the volumetric weight of slope deposits was considered 1.61, value determinated by Surdeanu (1998). For the length of slopes in direct contact with the bed, the lake area was taken into account, where sediments produced by creep enter directly into the lake, and for the basin, the length of the hydrographic network of 1st Order (Horton-Strahler classification), which has no developed bed and within and in which the slopes are in direct contact with it. The results show a total amount of sediments produced by this process of approx. 2493.7 m^3/year (in 23 years—57,355.1 m^3).

The river bank erosion was estimated using the total length of the hydrographic network for basins of 2nd and 3rd Order (Horton-Strahler classification), and the production of sediments given by one linear meter of shore was calculated using the formula proposed by Rădoane (2002) (Eq. 6.4).

Equation (6.4), The formula of production of sediments given by one linear meter of shore proposed by Rădoane (2002):

$$\log \mathrm{V}_{\mathrm{erodat}} = -1.4442 + 1.423 * \log \mathrm{H} \tag{6.4}$$

where: V—eroded volume; H—shore height.

Results of sediment volumes were obtained for basins of each order (1st–4th), within the basin of Cuejdel Lake, the final volume being obtained by taking into consideration the following effluence ratio proposed by Ichim et al. (1998): 1st Order—100%, 2nd Order—65%, 3rd Order—42%, 4th Order—33%. This equation was used in this study by applying it to obtain the production of sediments closest to 191.8 m^3/year (in 23 years—4411.4 m^3).

Regarding the forest roads erosion, this can be estimated by multiplying the average sediments production rate per meter multiplied by their length. Within the basin of Cuejdel Lake, the length of operating forest roads is 27.25 km, with an area of 817.71 ha (considering their average width of 3 m), but most of them with temporary use. According to Elliot et al. (1999), the authors estimate a sediment production rate below 60 t/ha/year for roads with slopes below 12% and lengths of less than 50 m. The average value of the slope of operating roads in the catchment of lake is 24.16%, but the flow length is reduced, generally below 20 m, and their utilization rate is also low. On these grounds, we estimate that an erosion of 2 t/ha/year is similar to the reality in the field.

In conclusion, based on the methodology of alluvial budget valuation for the Cuejdel catchment basin, the sediment budget evaluation of the lake could be done. Considering the effluence ratios of tributaries of various orders and of Cuejdel stream flowing into the lake, alluvial effluence can be estimated, being obtained the amount of sediments reaching the lacustrine basin. The result indicates for the catchment area an alluvial production of 3288.5 m^3/year (3.02 t/ha/year), in 23 years reaching cca. 75,635.93 m^3.

6.2 Sediment Volume Evaluation and Clogging Rate

Validation of the evaluation of the volume of sediments transported into the lake was achieved by direct methods (geophysical) of mapping alluvial deposits. Initially, the sediment thickness has been determined using a ground-penetrating radar device and a GPS for spatial georeference of the clogging model. The device used for scanning the lacustrine basin could operate without interruption for several hours. Alluvial structures displayed in real time on the device screen allowed in situ quality assessment. The second step consisted of filtering the GPR images in order to improve the reflection of high-frequency electromagnetic pulse. The internal structures of alluvial deposits, such as the current bottom of the lake, the original bottom of flooded riverbed, and sediment thickness were transferred as spatial database in GIS software sites and analysed by specific techniques (Mihu-Pintilie et al. 2014a).

The operation principle of the GPR device is the propagation of electromagnetic waves in alluvial deposits which highlights the division of layers (Conyers 2004; Reynolds 2011; Nicu 2013). The best results were obtained on a plain and dry topographic surface, and the depth of investigation had to be adapted to: relative permittivity (ε_r), permittivity of free space (ε_0) calculated according to 8.8542×10^{-12} F/m, and to antenna frequency selected for scanning, in our case 100 MHz. The measurements made within Cuejdel Lake took into account the values of relative permittivity of materials specific to aquatic environment.

Until now, this method was used mostly in archaeological survey, and more rarely in hydro-geological and geomorphological research. However, the GPR device was used for scanning Quaternary deposits (Derald and Harry 1995; Smith and Jol 1995; Bristow and Jol 2003; Bryan and Jeffrey 2006; Ndiaye et al. 2014), alluvial deposits from floodplains (Sass et al. 2007; McClymont et al. 2010), lacustrine sediments (Moorman and Michel 1997; Schwamborn et al. 2002; Charlie and Harry 2003; Banks and Johnson 2011; Stoleriu et al. 2015), coastal morphostructures (Buynevich and Fitzgerald 2003), island formation (Denziman et al. 2010), ponds (Shuman and Donnelly 2006), quagmires (Plado et al. 2011; Proulx-Mclnnis et al. 2013), soil water dynamics (Pan et al. 2012; Dagenbach et al. 2013), glacier morphostructure (Gabbi et al. 2012; Monnier et al. 2014).

The measurements started in the winter of 2013 and resumed in February 2014. The thickness of the ice bridge ranges between 25 and 45 cm. The scan was based on 12 cross sectional profiles with lengths between 72 and 228 m and a longitudinal profile 915 m long. The longitudinal profile tracked the path of the old water course from upstream to downstream, and intersected all cross-sections. GPR images were georeferenced in the stereographic projection of Romania and the correction was

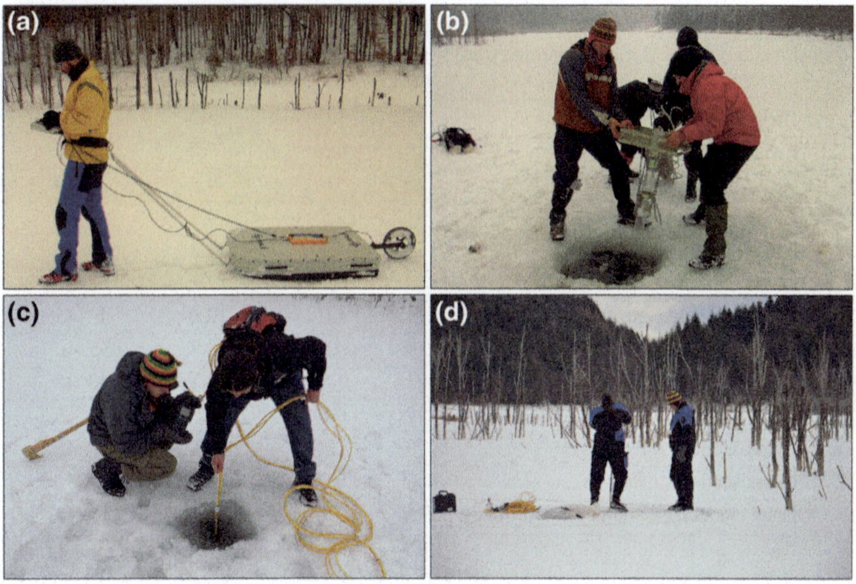

Fig. 6.3 Collecting data in February 2013: **a** GPR surveys with 100 MHz shielded antenna, **b** Extracting sediment samples using a Free-fall corer, **c** Analysis of the physical–chemical characteristics of sedimentary environment for calibrating the GPR device, **d** Measurement of the ice bridge and water depth

made using GPS spatial data to remove cartographic projection errors (Mihu-Pintilie et al. 2014a) (Fig. 6.3; Table 6.5).

Data processing was based on the internal software of GPR device, a method which involved filtering the images according to the following operations: (A) dewow the data for all traces, 10 ns time window, (B) gain the data for all traces, energy decay, 1.0 scaling values, (C) static correction start time, −0.25 ns manual, (D) band pass filter different for each cross section and (E) removal the background (Buynevich and Fitzgerald 2003; Charlie and Harry 2003; Denziman et al. 2010; Plado et al. 2011; Banks and Johnson 2011).

GPR images allowed mapping the current and initial bathymetric profile, the thickness of lacustrine deposits, the shape of initial valley, old river depositions, the dam structure and bedrock (Figs. 6.4, 6.5 and 6.6). In some cases we had to remove a number of scanning errors due to submersible trees that interfered with electromagnetic propagation wave, thus producing diffractions on the reading interface of signal (Fig. 6.7).

The thickness of lacustrine deposits ranges between 0.1 and 3.6 m with an average of 0.8–1.0 m. Thickness values over 3.0 m correspond to cross-sectional profiles No. 4—3.2 m, No. 9—3.7 m, No. 11—3.51 m, No. 12—3.4 m and No. 13—3.42 m. The classes with medium thickness between 1 and 2 m have shown a high frequency in

Table 6.5 Characteristics of GPR profiles (Mihu-Pintilie et al. 2016)

No.	Profile type	Orientation/direction	Length (m)	Min. depth (m)	Max. depth (m)	Ice bridge (cm)
1.	Cross section	North–South	92	1.0	3.2	15–35
2.	Cross section	East–West	127	0.5	7.5	25–35
3.	Cross section	East–West	120	0.5	9.8	30–40
4.	Cross section	East–West	91	2.8	9.5	40–45
5.	Cross section	East–West	228	1.0	11.3	40–45
6.	Cross section	Northeast–Southwest	179	0.8	12.2	35–40
7.	Cross section	Northeast–South	211	1.0	12.0	40–45
8.	Cross section	Northeast–Southwest	219	0.8	13.8	35–40
9.	Cross section	Northeast–Southwest	73	1.0	15.1	40–45
10.	Cross section	Northeast–Southwest	72	0.8	14.9	40–45
11.	Cross section	East–West	79	1.5	12.6	40–45
12.	Cross section	East–West	63	0.0	3.5	25–35
13.	Long. section	Southeast–Northwest	914	0.2	15.1	15–45

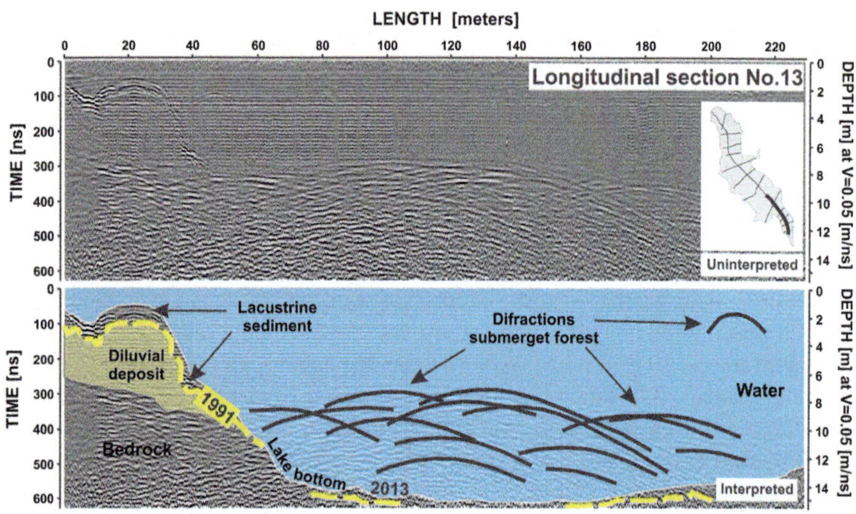

Fig. 6.4 GPR image: Longitudinal section No. 13 in the natural dam area

the middle of cross sectional profiles which indicate a constant progradation along the old water course of Cuejdel brook (Table 6.6).

For data validation, 4 sedimentary core samples (Free-fall corer) were taken from the intersection of No. 5, No. 8 and No. 10 cross sections with the longitudinal profile No. 13 (core dimension: L—1 m; Ø—6.5 cm). The choice of their location was determined on the following considerations: core sample No. 1—the presence of alluvial cone from the upper lake basin; core sample No. 2—the maximum erosion zone in line

Table 6.6 Static conductivity and relative permittivity values of fresh water and sedimentary environments and an example of interpreting a GPR image (Buynevich and Fitzgerald 2003; Charlie and Harry 2003; Denziman et al. 2010; Plado et al. 2011; Banks and Johnson 2011; Mihu-Pintilie et al. 2016)

			Static conductivity O'$_s$ [mS/m]		Relative permittivity ε_{ave}		
No.	Material	Type	Min.	Max.	Min.	Max.	GPR image
	Air	-	0	-	1	-	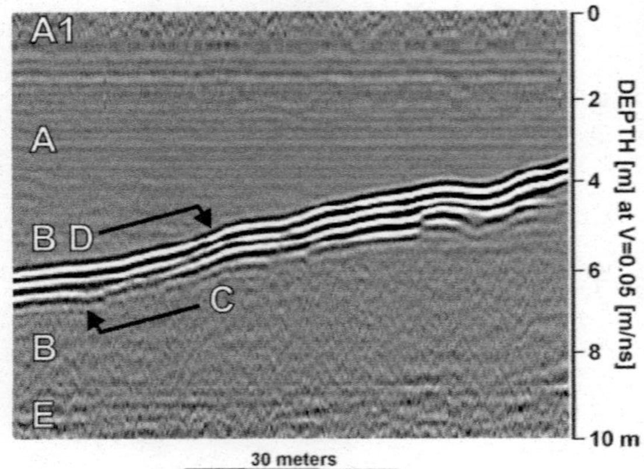
A1	Fresh water	Frozen	1	10^{-5}	3	-	
A		Liquid	10^{-1}	10	78	88	
B	Sand	Dry	10^{-3}	1	3	6	
		Wet	0.1	10	10	30	
C	Soil	Dry sandy	0.1	100	4	6	
		Wet sandy	10	100	15	30	
		Dry clayish	0.1	1	4	6	
		Wet clayish	10	100	10	20	
		Dry loamy	10^{-1}	100	4	6	
		Wet loamy	100	1000	10	15	
		Normal	5	-	16	-	
D	Clay	Dry	1	100	2	20	
		Wet	100	1000	15	40	
E	Limestone	Dry	10^{-2}	10^{-6}	4	8	
		Wet	10	100	6	15	
	Sandstone	Dry	10^{-2}	10^{-6}	4	7	
		Wet	10^{-1}	10^{-2}	5	15	
	Schists	Saturated	10	100	6	9	

Table 6.7 Morphometric characteristics of sediment samples taken with gravity core from Cuejdel Lake (Mihu-Pintilie et al. 2016)

No./type	Water depth (m)	Ice bridge (cm)	Core sample thickness (cm)	No. of lamination	No. of sample
1. Gravity core	3.50	45	39.00	15	35
2. Gravity core	11.50	40	49.00	25	35
3. Gravity core	14.90	35	47.00	30	40
4. Gravity core	8.50	15.9	48.00	15	45

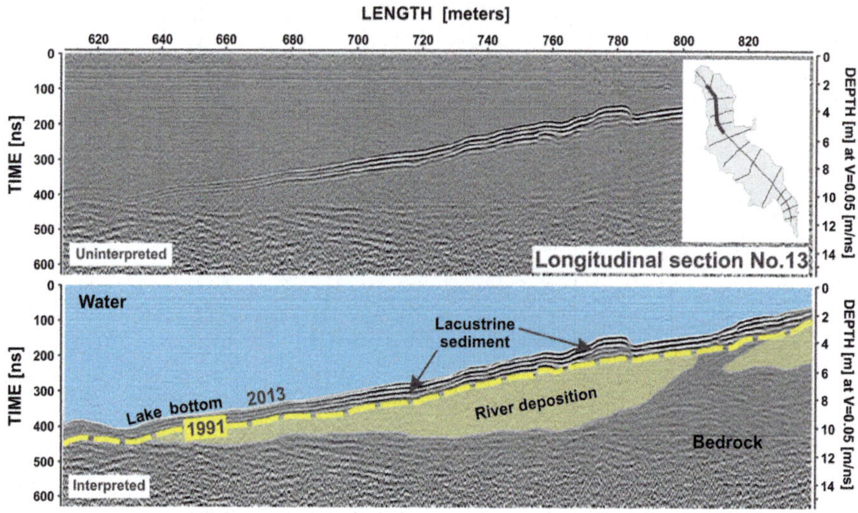

Fig. 6.5 GPR image: Longitudinal section No. 13 in the main alluvial fan area

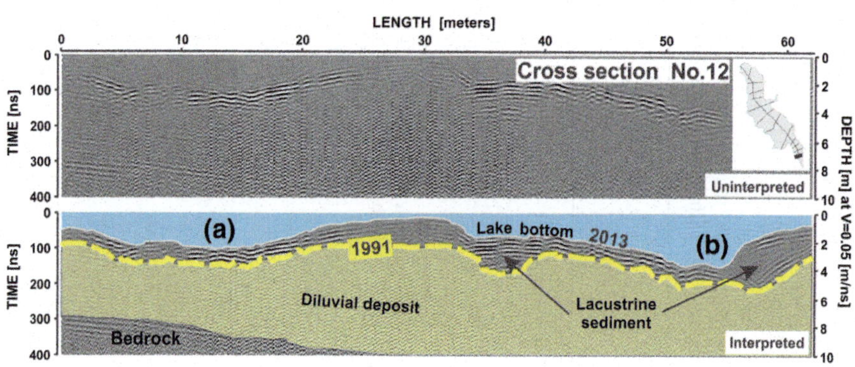

Fig. 6.6 GPR image: Cross section No. 12 with the indication of first (**a**) and current (**b**) outlet

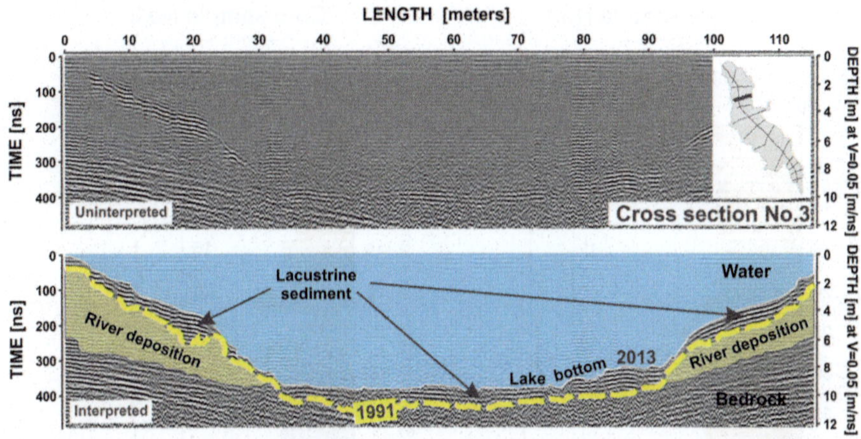

Fig. 6.7 GPR image: Cross section No. 3 in the main alluvial fan area

Table 6.8 Values of sediment budget and differences between indirect (USLE) and direct (GPR) methods for estimate the clogging rate of Cuejdel Lake

Method	Erosion (t/ha/year)	Total volume (m^3)	Annual average (m^3/year)
USLE (indirect method)	3.027	75,635.93	3288.51
GPR measurements	4.334	103,988.87	4521.26
Differences	1.307	28,352.94	1232.68

with the tributary Făgetu; core sample No. 3—the max. depth area of lake basin; core sample No. 4—the presence of alluvial cone from the dam area (Fig. 6.8; Table 6.7).

The thickness of extracted samples ranged from 54 to 72.5 cm wet, but the analysis of laminations was possible only after extruding the samples in semi-dry state, which led to the compaction of approx. 5–10 cm/sample. For this reason, the granulometric analysis of the laminas follows only a description of sedimentation conditions and a confirmation of alluvial deposits thickness determined by direct measurement. Meanwhile, from the four samples only core sample No. 2 surprised the whole package of sediments by reaching the initial soil layer covered by alluvium.

For each sample were identified between 16 and 21 successive laminations, their thickness ranging from 1 mm to 20 cm. At first glance, the wide diversity of sedimentary load for each representative area of the lake does not allow a generalization of processes and products in a simple scheme. This is influenced by the terrigenous intake and by the nature of the different source-area (surface erosion, creep, lake bank erosion, in situ settling, biodegradation of organic matter, etc.), by the variation in the depth of the lake where samples were collected, and by precipitation, temperature, hydrodynamic regime of water in different sectors, biotic activity, etc. However, all samples show a good horizontal stratification and frequent changes of color vertically depending on the lithological nature of the sediments.

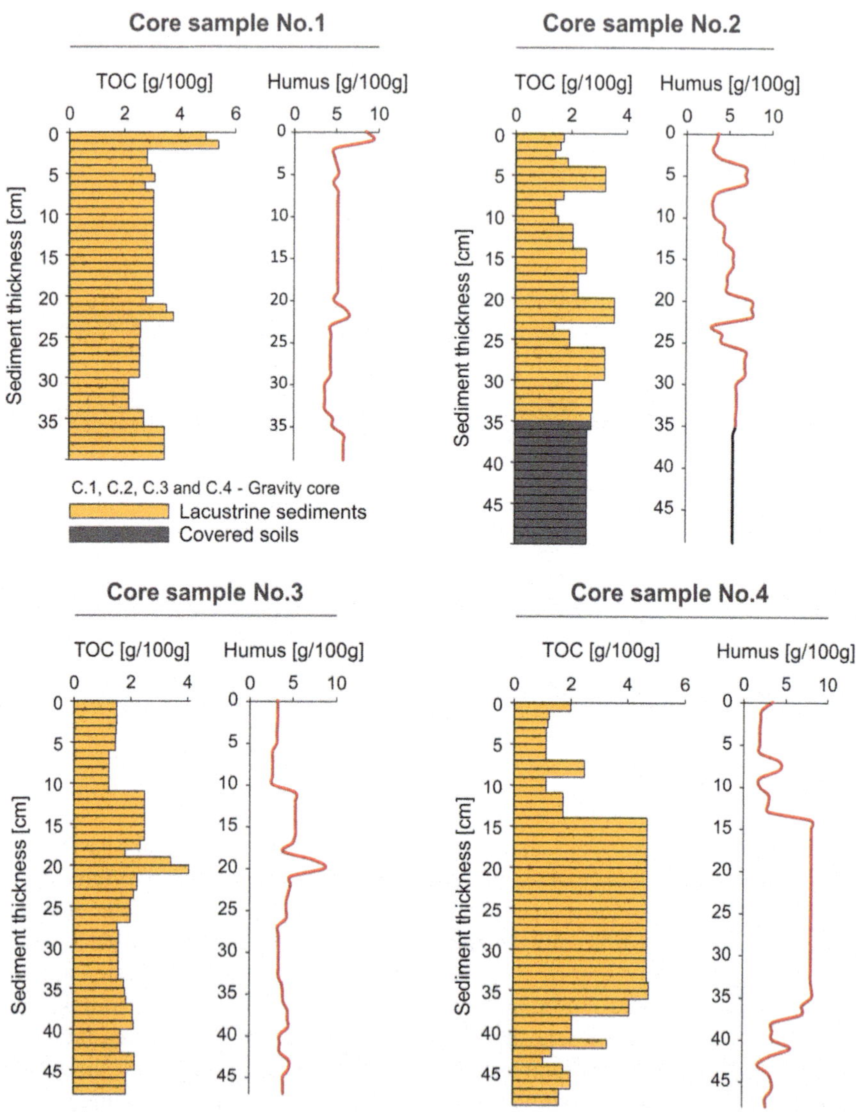

Fig. 6.8 Vertical distribution of total organic carbon (TOC—g/100 g) and humus (g/100 g) in the sediment samples from Cuejdel Lake

In each tested area, lithological lacustrine sequences present freshwater limestone with thin dolomitic interlayers (1–3 mm), bituminous clays (0.3–20 cm), iron oxides, concavities with ferruginous and sideritic evaporites auras (Ø 1–5 mm), and fossils belong to a small number of species, mostly frustules of diatoms. Most annual nonglacial varves are a consequence of dimictic character of the lake, sedimentation being influenced by thermal stratification (Mihu-Pintilie et al. 2014b, c). However, lack of dissolved oxygen content and of nutrients from the hypolimnion zone determines the accumulation of characteristic sediments of seekreide type (sediments composed of bioturbated marly silts, silty clays, clays, marls and carbonate muds), gyttja (strongly bioturbated marly silts) and sapropel (fine mud rich in organic detritus) (Mihu-Pintilie et al. 2014b).

The qualitative analysis of lacustrine sediments from Cuejdel Lake was focused on identifying the granulometric ratio of detrital sediments (clay/silt/sand) and of total organic carbon content. Thus, in the sample taken from the alluvial cone from the upper lake basin (core sample No. 1), the first 20 cm indicate a high content of fine clays ($Ø_{med}$ < 0.004 mm), mixed with organic matter (abundant between 7–9.5 cm) and the total organic carbon ranges between 2.7 and 5.4 g/100 g (humus—4.7 to 9.3 g/100 g). Silty marls from the upper parts of deposits ($Ø_{med}$—0.004 to 0.063 mm) indicate a dynamic process of accumulation of sediments alternating with biodegradation of organic matter, but which is subject to quantity. Siltic stratification occurs at 20–25 cm, mixed with organic matter ranging from 2.5 to 3.5 g/100 g (humus—4.4 to 6.5 g/100 g). At the sample basis, the lamination presents a higher content of fine sand and silt, the amount of total organic carbon decreasing to 35 cm, below 2.5 g/100 g. Last laminations between 35–39 cm, indicate an increase of 1 g/100 g of total organic carbon (TOC—3.4 g/100 g; humus—5.9 g/100 g), but this time in a finer clayey matrix.

The annual nonglacial varves from the core sample No.1 indicate by fine layers of micritic crystals of calcite a chemical precipitation of $CaCO_2$ in summer, and by the thicker laminas of siltic material, a specific accumulation of periods with heavy rainfall and a supercritical flow of Glodu and Cuejdel streams. Within the layer of 7–9.5 cm, undecomposed organic matter (leaves, wood fragments) certifies an event of this type. Generally, the sedimentation rate from the upper part of the alluvial cone in the upper lake basin demonstrates a high solid transportation of the two tributaries.

The sample taken from the central area of the lake (core sample No. 2), respectively from the outlet area of the Făgetu stream, surprised the entire column of lake sediments reaching the submerged soil layer. Within the first 26 cm laminas of seekreide type dominate ($Ø_{med}$—0.002 to 0.063 mm), showing fine successions of marly silts with thin dolomitic (1 mm) or sapropelic (2–3 mm) interlayers. Concavities with ferruginous and sideritic auras (Ø 2–4 mm) have a high frequency. The amount of organic carbon has an average value of 2.07 g/100 g (TOC_{min}—1.3 g/100 g; TOC_{max}—3.5 g/100 g). Among the range of 26–37 cm, the average size of particles exceeds Ø 0.063 mm, and the amount of total organic carbon ranges between 2.7 and 3.2 g/100 g. At this depth, transition occurs between lacustrine sediments and upper part of sunken soil, rich in undecomposed organic matter and fine arenitic materials (Ø 0.3–0.5 mm). From this depth to the base (49 cm), the sample presents the characteristics of a typical soil composed of undegraded organic matter (leaves, straw, plant roots, pieces of wood, bark), with a compact structure without laminations.

Reduced thickness of lacustrine sediments in this sector (35 cm) confirmed by GPR measurements shows a strong erosion activity of Făgetu tributary. The large declivity of the submerged valley induces a high rate of penetration of rain water with solid load, its storage taking place to the lake outlet. For this reason, the sector is characterized by fine alluvial deposits which are cemented in dry periods by carbonate precipitation. Sapropelic laminas have reduced thickness (<2 mm) due to the removal of organic matter on the bottom lake in periods of high turbidity.

In the core samples No. 3 and No. 4 extracted from areas with maximum depths of the lake (>14 m), the lacustrine sediment lamination presents a series of correspondences. The core sample No. 3 describes a series of arrhythmic processes of accumulation, where intervals with deposits of fat clays and silty clays ($Ø_{med}$ < 0.004 mm) overlap sequentially over thick layers of fine sands ($Ø_{med}$—0.3 to 0.4 mm). Lenses of sapropel rich in organic detritus (TOC—3.4 to 4.0 g/100 g; humus—5.8 to 6.9 g/100 g) occur in the interval 10.5–11 cm, respectively 19–21 cm, in the upper part of the thick laminas of fine sand. The core sample No. 4 drawn off at approx. 100 m away of core sample No. 3 to the dam presents within the interval 14–34 cm compact and without stratification fine arenites ($Ø_{med}$—0.3 to 0.7 mm). These laminas are similar to those identified in the previous sample (layers 11–17 cm and 21–23 cm), both concerning the granulometric constitution and the amount of total organic carbon (TOC—4.08 to 4.7 g/100 g; humus—7.04 to 8.2 g/100 g).

The correlation between the two consistent laminations of fine sand identified in the core samples No. 3 and No. 4, with a similar load of TOC, as well as the same black sapropel lens from the upper part, rich in organic detritus, most likely indicate the expression of the same morphological and hydrodynamic event. According to the depth where laminations were identified, respectively in the first 35 cm, the hypothesis can be issued, that these deposits are a consequence of flood produced on 12/VII/2005, when in Cuejdiu River basin cumulative rainfall only in few hours were 79.9 mm, resulting a flood at Cuejdiu station with a maximum flow rate of 32.9 m^3/s, which is a rate of approx. 60 higher than the average (Q_{med}—0.54 m^3/s).

Under these sedimentation conditions, after interpolating the bathymetric information collected from both GPR profiles and gravity cores, using software specific to GIS (ArcGIS—Topo to Raster, TNTmips), the results were concretized in achieving two Digital Elevation Models (DEM) of the lacustrine basin. The first bathymetric model corresponds to the initial configuration, respectively the flooded valley sector. The second model shows the current shape of the lacustrine basin, in this case using as reference the bathymetric map created with the sonar in 2011. The difference between the two models is basically the thickness of sediments accumulated in Cuejdel Lake since its genesis and until 2013 (Fig. 6.9).

The distribution model of lacustrine sediments in Cuejdel Lake indicates that the sections with thickness of alluvial deposits between 0.1 and 1.0 m cover 8.5 ha (60.7%), the range 1.0–2.0 m approx. 4.6 ha (33.32%), and the range >2.01 m the rest of 8326.5 m^2 (5.93%). From the most representative interval, the sediment thickness ranging between 0.8 and 1.0 m covers the largest surface, respectively 2.5 ha, followed by the range 0.6–0.8 m with 2.9 ha and 1.0–1.2 m with 2.04 ha. In total, the range between 0.4 and 1.4 m covers approx. 9.9 ha (71.16%), while the extremely thick deposits have a low distribution (<0.2 ha) (Fig. 6.10).

The areas with >1.0 m alluvium thickness have a higher frequency in the shoreline corresponding to the right slope, between the outlet of Pârâul Rotund and Glodu

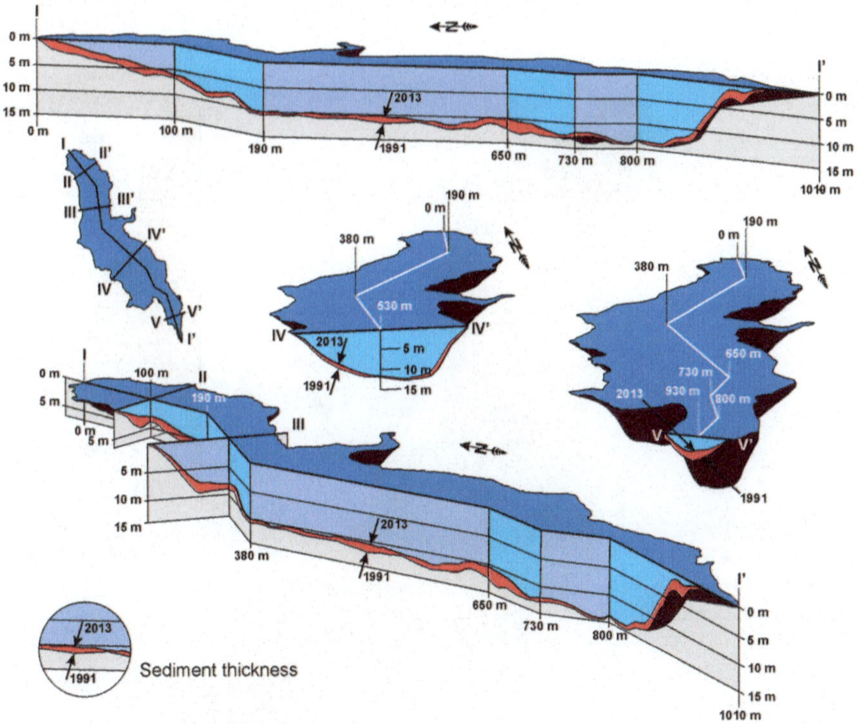

Fig. 6.9 3D sections across the Cuejdel lacustrine basin (Mihu-Pintilie et al. 2016)

brooks. The low value of sediment deposits is due to submerged slopes with declivity between 5° and 15° and high density of submerged fossil trunks. Also, the western shoreline, fully covered by forest, and the low fragmentation degree of the slope, induce a low erosional activity.

At the outlet of Piciorul Crucii brook, the alluvial deposits are transported to the old confluence with Cuejdel brook, especially when the torential activity is increased. An almost identical situation occurs on the left slope of the lake, between the outlet of Făgetu, Cuejdel and Glodu brooks. In general, the lake slopes present reduced lengths which determines, due to level oscillations, the transport of sediments to a deeper area. The average value of accumulation ranges between 2.1 and 4.3 cm/year.

Low sediment thickness is also found in Piciorul Crucii brook, with a range between 0.4 and 1.0 m and Făgetu brook with a range between 0.1 and 0.5 m. These thicknesses indicates an high erosional activity. The valleys of the two tributaries show high declivity, with range between 15° and 18°, continuing to the center of the lake trough submerged slopes >35°. For this reason, the sediments are transported by water dynamics and deposited in the old flooded riverbed of Cuejdel stream. At the old confluence of Făgetu and Cuejdel brooks, where the main current receives the secondary currents, the high turbidity forms small circular depressions where

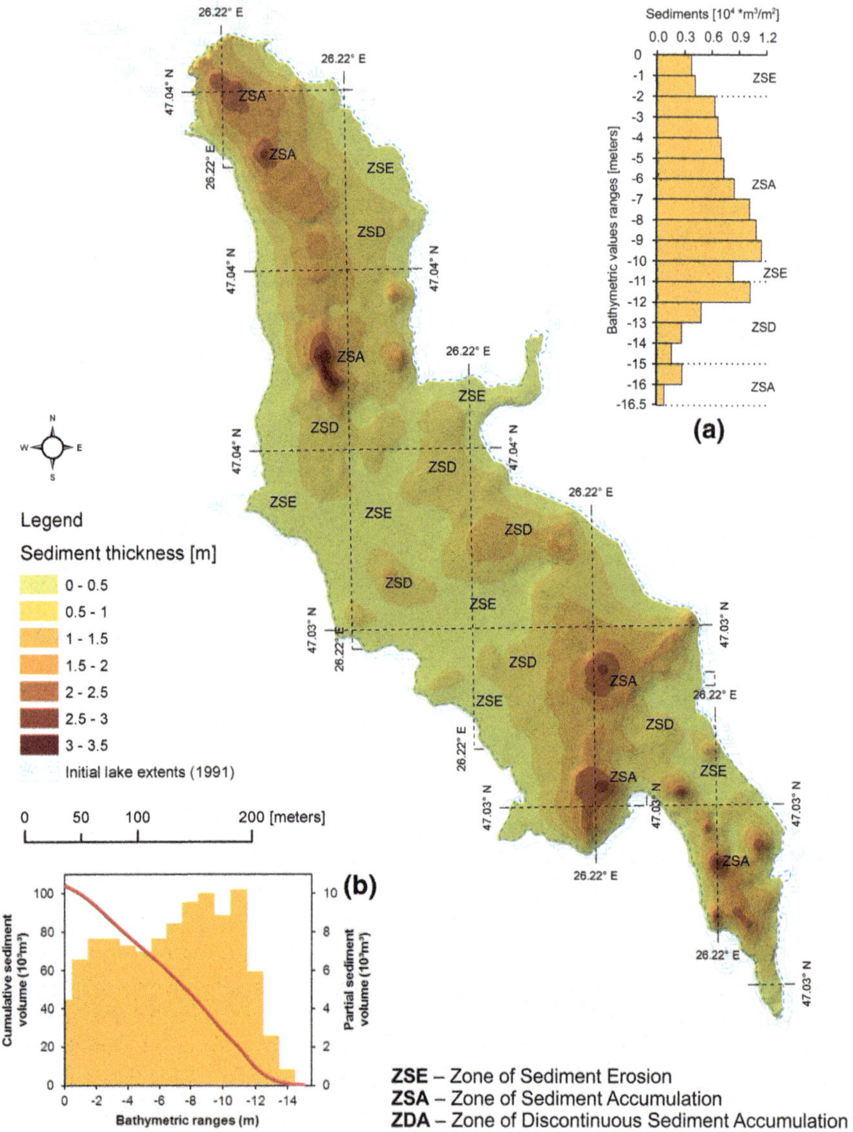

Fig. 6.10 Sediment thickness map of Cuejdel Lake: **a** Delimitation of erosion and accumulation areas, **b** Partial and cumulative sediment volume (Mihu-Pintilie et al. 2016)

the thickness of the sediments is >0.2 m, confirmed by the gravity core No. 2. The average clogging rate is <1 cm/year (Mihu-Pintilie et al. 2014a, b).

The areas with the highest lacustrine sediments thickness have been measured in the upper sector of the lake, along the north-western slope of Piciorul Rotund, in the area with maximum depths and also on the dam surface. The alluviums brought into the lake through Cuejdel and Glodu streams form an alluvial fan in the upper sector of the lake with thickness values ranging between 1.5 and 3.6 m. The progradation occurs along the initial valley, close to the old confluence with Făgetu brook. The elongated form of deposits in the north-south direction indicate the lack of trees in the bottom of lacustrine basin. The average clogging rate from this sector ranges between 6.5 and 15.5 cm/year.

In the area with depths over −12 m and on the right slope of Pârâul Rotund stream, the deposit thickness is 1.8–3.2 m. The accumulation takes place as a consequence of torrential activity on the right slope. The cross-sectional shape of the alluvial bank, in south-east–north-west direction is dictated by the general currents direction and the slopes with declivity range between 35° and 50°. The average accumulation value of sediments in this area is 10 cm/year, reaching in the area with depths over −14 m a rate of 15 cm/year.

The submerged area in front of the natural dam has a very fragmented morphology, influenced by the chaotic distribution of deluvial materials on the lacustrine basin bottom. For this reason, the alluvial deposits show a variable thickness ranging between 0.4 and 3.6 m and the average annual rate of sedimentation reaches up to 16 cm/year. The high value of alluvial deposits thickness is induced by slope >45°, frequently dislocated by mud lobes deposited on the bottom of the lacustrine basin. Also, the solid load transported in the direction of the main current is deposited at the base of the natural dam forming a submerged bank (Fig. 6.10).

On average, the clogging rate of lacustrine sediments is between 1 and 16 cm/year and the extreme values correspond to reduced areas. The highest frequency ranges between 3.5 and 5 cm/year indicating a high percentage of alluvial deposits with a thickness of 0.8–1.15 m and accumulated in the last 23–24 years. On bathymetrical levels, the volume of alluvial deposits ranges between 33.6 and 40.4 m^3/year. The average volume of sediments accumulated on surfaces between 1 and 1.2 ha is of 300–450 m^3/year. The estimation of sediment volume was achieved based on the sum of all partial volumes resulted from the substraction of the two bathymetric models. Since 1991 and until the winter of 2014, in the lake have accumulated approx. 10.39×10^4 m^3 which represents approx. 8.5% of the lake volume in 1991. The annual clogging rate is around $0.45 \times 10^4 m^3$/year (0.36%/year), indicating a hypothetical clogging period of Cuejdel Lake of $\simeq$250 years.

The sediment amount estimated by GPR method is higher by $+2.83 \times 10^4$ m^3 ($+0.12 \times 10^4$ m^3/year) that the one determined by equations of erosion at the level of hydrographic basin. Thus, the erosion value in the catchment basin and the effluence rate of alluvium determined based on direct measurements is of 4.3 t/ha/year. The difference of 1.3 t/ha/year over the sediment budget determined trough indirect method (USLE) can be attributed to either underestimation of the sediment production values of a process, or to the failure to include existing processes at the level of the catchment basin. Landslides that occur in the minor riverbeds of 1st Order could be such a phenomenon (Table 6.8).

References

Arghiriade C (1977) Rolul hidrologic al pădurii. Ceres Publishing, Bucharest
Banks WSL, Johnson CD (2011) Collection, processing, and interpretation of ground-penetrating radar data to determine sediment thickness at selected location in deep Creek Lake, Garrett County, Maryland. US Geol Survey–Scientific Investigations Report 2011–5223, Reston, Virginia
Bristow CS, Jol HM (2003) An introduction to ground penetrating radar (GPR) in sediments. In: Bristow CS, Jol HM (eds) Ground penetrating radar in sediments. SP Geol Soc, London 211:1–7
Bryan S, Jeffrey PD (2006) The influence of seasonal precipitation and temperature regimes on lake levels in the northeastern United States during the Holocene. Quatern Res 65:44–56
Buynevich IV, Fitzgerald DM (2003) High-resolution subsurface (GPR) imaging and sedimentology of coastal ponds, Maine, USA: implications for Holocene back-barrier evolution. J Sediment Res 73(4):559–571
Charlie SB, Harry MJ (2003) An introduction to ground penetrating radar (GPR) in sediments. Geol Soc Bull, SP London 211:1–7
Conyers LB (2004) Ground-penetrating radar for archaeology. Alta Mira Press, Walnut Creek
Dagenbach A, Buchner JS, Klenk P, Roth K (2013) Identifying a parameterisation of the soil water retention curve from on-ground GPR measurements. Hydrol Earth Syst Sci 17:611–618
Denziman C, Brevik EC, Doolittle J (2010) Ground-penetrating radar investigation of a rapidly developed small island in a lake in southern Georgia, USA. J Cave and Karst Stud 72(2):94–99
Derald GS, Harry MJ (1995) Ground penetrating radar: antenna frequencies and maximum probable depths of penetration in quaternary sediments. J App Geophys 33:93–100
Diaconu C (1992) Probleme de metodica determinării curbelor teoretice ale probabilităţilor de depăşire (asigurare) în hidrologie. Hidrotehnica–Extras 37(4–5), Bucharest
Elliot JG, Gellis AC, Aby SC (1999) Evolution of arroyos: incised channel of the southwestern United States. In: Simon A (ed) Darby SE. Incised River Channels, Wiley, pp 153–185
Gabbi J, Farinotti D, Bauder A, Maurer H (2012) Ice volume distribution and implications on runoff projections in a glacierized catchment. Hydrol Earth Syst Sci 16:4543–4556
Ichim I (1986) Sistemul aluviunilor. In: Received of 1st Symposium–Provenienţa şi efluenţa aluviunilor, 5–96 November 1986, 1:1–31, Piatra Neamţ
Ichim I, Rădoane N, Rădoane M, Grasu C, Miclăuş C (1998) Dinamica sedimentelor. Aplicaţie la râul Putna-Vrancea. Technical Publishing, Bucharest
McClymont AF, Hayashi M, Bentley LR, Muir D, Ernst E (2010) Groundwater flow and storage within an alpine meadow-talus complex. Hydrol Earth Syst Sci 14:859–872
Mihu–Pintilie A, Romanescu G (2011) Morphometric and morphological suitability of the relief from the Crucii (Cuejdel) Lake basin (Stânişoarei Mountains). In: Pandi G, Moldovan F (eds) Proceedings of the air and water components of the environment, Babeș–Bolyai University, Cluj, 18–19 March 2011, vol 1, pp 305–313
Mihu-Pintilie A, Romanescu G, Stoleriu CC, Nicu IC, Asăndulesei A, Schmaltz E (2014a) Natural dam lakes from Cuejdiu watershed (Stânișoarei Mountains)—Non–invasive methods used for bathymetric maps. In: In: Gâştescu P, Marszelewski W, Breţcan P (eds) Proceedings conference of the water resources and wetlands, Tulcea, 11–13 September 2014, vol 1, pp 130–137
Mihu-Pintilie A, Romanescu G, Stoleriu C (2014b) The seasonal changes of the temperature, pH and dissolved oxygen in the Cuejdel Lake, Romania. Carpat J Earth Environ Sci 9(2):113–123
Mihu-Pintilie A, Paiu M, Breabăn IG, Romanescu G (2014c) Status of water quality in Cuejdi hydrographic basin from Eastern Carpathian, Romania. In: SGEM–proceedings of the 14th international multidisciplinary scientific geoconferences–hydrology and water resources, 18–24 June, Albena, vol 14, no 1, pp 639–646
Mihu-Pintilie A, Asăndulesei A, Nicu IC, Stoleriu CC, Romanescu G (2016) Using GPR for assessing the volume of sediments from the largest natural dam lake of the Eastern Carpathians: Cuejdel Lake, Romania. Environ Earth Sci 75:710. https://doi.org/10.1007/s12665-016-5537-1

Monnier S, Kinnard C, Surazakov A, Bossy W (2014) Geomorphology, internal structure, and successive development of a glacier foreland in the semiarid Chilean Andes (Cerro Tapado, upper Elqui Valley, 30° 08′S, 69° 55′W). Geomorphology 207:126–140
Moorman BJ, Michel FA (1997) Bathymetric mapping and sub-bottom profiling through lake ice with ground-penetrating radar. J Paleolimnology 18:61–73
Moțoc M, Stănescu P, Taloescu I (1979) Actual conceptions regarding erosional phenomenon and it control. Inst Soil Sci Agrochemistry, Agriculture Library, Bucharest
Ndiaye M, Clerc N, Gorin G, Girardclos S, Fiore J (2014) Lake Neuchâtel (Switzerland) seismic stratigraphic record points to the simultaneous Würmian deglaciation of the Rhône Glacier and Jura Ice Cap. Quat Sci Rev 85:1–19
Nicu IC (2013) Analiza riscurilor hidrogeomorfologice care afectează siturile arheologice eneolitice din bazinul hidrografic Valea Oii (Bahlui)–Studii de caz. Alexandru Ioan Cuza University of Iaşi, Disertation
Pan X, Zhang J, Huang P, Roth K (2012) Estimating field-scale soil water dynamics at a heterogeneous site using multi-channel GPR. Hydrol Earth Syst Sci 16:4361–4372
Plado J, Sibul I, Mustasaar M, Jõeleht A (2011) Ground-penetrating radar study of the Rahivere peat bog, eastern Estonian. Estonian J Earth Sci 60:31–42
Proulx-McInnis S, St-Hilaire A, Rousseau AN, Jutras S (2013) A review of ground-penetrating radar studies related to peatland stratigraphy with a case study on the determination of peat thickness in a northern boreal fen in Quebec, Canada. Progress in Physical Geography 37(6):767–786
Rădoane N (2002) Un nou lac de baraj natural în bazinul Bistriţei Moldoveneşti-Lacul Cuejdel. Stud Cerc Geograph 49:76–82
Reynolds JM (2011) An Introduction to Applied and Environmental Geophysics–2nd Edition. Wiley Blackwell
Sass O, Krautblatter M, Morche D (2007) Rapid lake infill following major rockfall (bergsturz) events revealed by ground-penetrating radar (GPR) measurements, Reintal, German Alps. The Holocene 17(7):965–976
Schwamborn GJ, Dix JK, Bull JM, Rachold V (2002) High-resolution seismic and ground penetrating radar-geophysical profiling of a Thermokarst Lake in the Western Lena Delta, Northern Siberia. Permafrost Periglac Process 13:259–269
Shuman B, Donnelly JP (2006) The influence of seasonal precipitation and temperature regimes on lake levels in the northeastern United States during the Holocene. Quat Res 65:44–56
Smith DG, Jol HM (1995) Ground penetrating radar: antenna frequencies and maximum probable depths of penetration in quaternary sediments. J Appl Geophys 33:93–100
Stoleriu CC, Romanescu G, Romanescu AM, Mihu-Pintilie A (2015) Morpho-bathymetrical conditions and the silting rate in Stanca-Costesti reservoir (Romania). Wulfenia J 22(2):451–470
Syvitski JPM, Milliman J (2007) Geology, geography, and humans battle for dominance over the delivery of fluvial sediment to the coastal ocean. J Geology 115:1–19
Surdeanu V (1998) Geografia terenurilor degradate–Alunecări de teren. Cluj University Press, Cluj-Napoca
Vanmaercke M, Obreja F, Poesen J (2014) Seismic control on contemporary sediment export in the Siret river catchment, Romania. Geomorphology 216:247–262

Chapter 7
The Seasonal Variation of Physical-Chemical Parameters and Water Quality Assessment (WQI)

Abstract The physico-chemical analysis of Cuejdi hydrographic basin and the Cuejdel lake basin was realized based on a monitoring program which consisted of analysing the seasonal variation of main physico-chemical parameters, with major implications in establishing the degree of trophicity in which the lacustrine ecosystem is located and also regarding the qualitative identification of the biotic component. The scientific approach consisted of two distinct methods: measurements in situ and sampling evidences for laboratory research. In situ, the measurements were made with a portable multi-parameter (Hach Lange device) to collect physico-chemical data. Based on this, Cuejdel Lake acts as a dimictic reservoir with direct thermal stratification during the summer, spring and autumn, and reverse stratification during the winter. At the lowest depths, the temperature is constant all year (5 °C). The surface water is influenced by the air temperature. The water is slightly alkaline with values that oscillate around 8. The alkalinity of the water is caused by the elements that are transported in the water as the hydrographic basin drains. The water in the upper section of the lake is very rich in oxygen. Below 10–11 m, dissolved oxygen is not present. The assessment of water quality from Cuejdi River was performed through 13 parameters. 30 samples were collected during spring and autumn, measurements being made both in situ aquatic ecosystem (Multi 350i/SET WTW) and in the laboratory (bench meter Hanna HI 4421, Titroline 700 SI Analytics and WTW Turb 555IR). The main fact is that a water quality parameter has no equal contribution to the final stage of water quality in all seasons; they can be different from both natural and anthropogenic causes. The analysis, especially the spatial distribution of the nitrate levels, revealed an important degradation of the water quality owed to the anthropogenic pressure. Furthermore, the findings confirm that the natural contamination of the Cuejdi River is less harmful than the contamination with compounds resulted from anthropogenic activities.

In situ measurements were made using a Hack Lange multiparameter instrument consisting of four sensors for main physico-chemical parameters of water with real time display: temperature (°C), pH (pH units), Luminiscent Dissolved Oxygen (LDO101) and Standard Conductivity (CDC401). The supervision was seasonal (2011–2014)

A. Mihu-Pintilie, *Natural Dam Lake Cuejdel in the Stânişoarei Mountains, Eastern Carpathians*, https://doi.org/10.1007/978-3-319-77213-4_7

Fig. 7.1 Measurements performed during different seasons to collect physico-chemical data for water quality using Nansen bottle (**a**) and Hack Lange Multiparameter (**b**)

in 7 sampling points on the lacustrine surface, closely monitoring the variation on 1 m high vertical bathymetrical levels (Fig. 7.1).

The second method consisted in laboratory analyses aiming to determine the water quality index (WQI) based on 30 recorded samples taken from Cuejdi River and Cuejdel Lake. In order to collect the samples, a bathymetrical instrument specially adapted for freshwater environments (Nansen bottle) was used, with water storage capacity of 1 L water/sample and bathymetrical autonomy of >30 m. For the samples collected to determine the biological spectrum of water, a netting to filter 10 L water/sample was used. In the laboratory the following devices were used: a multiparameter instrument (Multi 350i/SET WTW), a bench meter (Hanna Hi 4421—DO/BOD/SOUR), a Titroline (SI Analytics—700) and WTW Turb (555IR), ion-selective electrodes and a Spectrophotometer (Shimadzu UV/VIS UV1601).

7.1 The Seasonal Variations of Physical-Chemical Parameters Determined In Situ

The morpho-bathymetric, hydrological, climate parameters and water dynamics are the crucial ecological factors for the organisms living in the water mass or in the surrounding of the lacustrine basin. Nevertheless, in Romania, the studies that focus on physico-chemical properties of water or natural lakes were made only sporadically (Gâştescu 1971; Pandi and Magyari 2004; Rădoane 2002; Romanescu 2009a, b, c; Romanescu and Stoleriu 2010; Romanescu et al. 2015; Xantus and Xantus 1999; Stoleriu et al. 2014). Concerning the analysis of thermal stratification of waters from the natural barrages and their classification based on physico-chemical properties in the Eastern Carpathians areas two barrage natural bodies of water can be identified: Cuejdel Lake and Red Lake, both studied in detail within the Geoarchaeological Laboratory from Alexandru Ioan Cuza University from Iași (Enea et al. 2011, 2012; Mihu-Pintilie and Romanescu 2011; Mihu-Pintilie et al. 2012, 2014a, b, 2015;

Romanescu et al. 2012; Romanescu and Stoleriu 2014), using the specific methodology from the extensive professional literature (Newcombe and Dweyr 1949; Sweers 1968; Dake and Harleman 1969; Edingher 1970; Sundaram and Rehm 1972; Davis et al. 1990; Stauffer 1992; Malm and Zilitinkevich 1994; Maberly 1996; Moalla 1996; Wand et al. 1997; Chapman et al. 1998; Ravens et al. 2000; Moundiotiya et al. 2004; Pandi and Buzilă 2004; Tadesse et al. 2004; Aplay et al. 2006; Glazik et al. 2006; Thakur and Bais 2006; Lampert and Sommer 2007; Nolan and Brigham-Grettej 2007; Belevantser et al. 2008; Skowron 2008; Araoye 2009; Maslanka 2009; Genova et al. 2010; Patra et al. 2010; Rucinski et al. 2010; Wassenaar 2012; Bukreev 2013).

With regard to a complex image of limnoecological environment, 7 sampling areas and measurements sites were chosen covering the morphological conditions of lake basin: main rivers outlet, maximum depth of the lacustrine basin, area around the landslide body, etc. The on-site analysis of the physico-chemical parameters was made using a Hach-Lange multiparameter instrument with real time display, and the measurements were made for each sample from the surface of the lake to the bottom of the lacustrine depression on 1 m high vertical bathymetrical levels. The monitoring frequency was seasonal and was repeated in spring, summer, autumn and winter to clearly document the variation of studied parameters over the course of one year cycle (4 measurements/year between 2011 and 2014). Using this methodology, complete data was obtained in order to correctly analyse the variation of life conditions in water bodies.

At the same time, the sampling points were carefully chosen to record the variations produced by the turbidity of the stream that supplied the lake. The locations of the measurements points required the use of GPS, although the measurements could not be performed at exactly the same point every time, especially due to the existence of surface and more importantly deep currents that affected the boat stability; therefore the measurements do not always present the same bathymetric values, although the differences are expressed in centimeters. The samples were collected during the day, when the sky was clear, between 09:00 AM–17:00 PM (Mihu-Pintilie et al. 2014a).

When the measurements were made, the water level variation was small (30–50 cm), the lowest levels being recorded during the winter as a result of the ice bridge while the highest were recorded in spring, generated by the melting of the snow. The water transparency, measured with the Secchi disk, varied between 0.50 and 3.5 m, the average being constantly situated between 1.8 and 2 m. The minimum transparency is recorded after the periods with maximum pluviometric values in the warm season or after the spring thaw as a result of water turbidity caused by the increase of solid flows. The maximum transparency is recorded during winter or in some cold autumn, when the plankton mass and the aquatic vegetation are missing.

Regarding thermal values, pH and dissolved oxygen, the records were made on the whole vertical profile of the lacustrine cuvette, as the morphology of the lake allows a high variation of physico-chemical parameters of water. The permanent and temporary water streams produce a vertical and surface variation through the qualities of transported waters. Also, the slopes situated in the close proximity of the

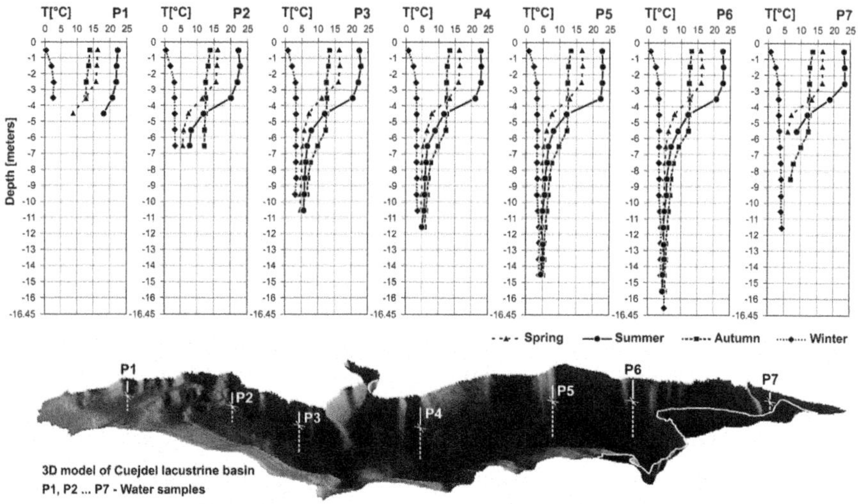

Fig. 7.2 Seasonal distributions of vertical thermal values (T °C) in the Cuejdel lacustrine basin (Mihu-Pintilie et al. 2014a)

lacustrine cuvette are covered with deciduous and coniferous trees and permit the development of wet areas only on small patches, usually until 70–100 cm.

At the water surface, the maximum temperature is almost 23 °C and is reached during the summer. The same temperature is maintained on the whole surface of the lake because the reduced flow of the rivers creates a very slow water dynamics. During the winter, when temperatures drop below 0 °C, an ice bridge is formed and can reach 25–50 cm thickness. As a result of the Cuejdel Valley features, the thermal inversions are frequently recorded which trigger air temperature decrease and the ice bridges at the lake surface can endure for a long period of time. In 2011, the ice layer formed at the surface of the lake had 37 cm thickness while in 2013 it measured 45 cm thickness. During the autumn a minimum temperature of 13 °C is recorded. The high value is explained by the fact that the large water mass hardly releases heat (Fig. 7.2).

At the maximum depth of 16.45 m, the temperature is constant, namely 5 °C. At the depth of 3 m, a sudden drop in temperature can be observed especially during spring and summer. During winter, at depths of over 2 m, the temperature constantly reaches 3–4 °C, and in autumn, the temperatures are relatively constant on the whole vertical profile until the depth of 5 m, where a sudden but slight drop takes place. In the outlet area of the Cuejdel River (upstream section) and in the dam area a disturbance in thermal values can be observed, with high differences in temperature between the four seasons and also at the maximum depths (P1, P2 and P7). Due to these thermal distributions, Cuejdel Lake represents a dimictic aquatorium. During summer and transitional seasons the water has a direct thermal stratification when the water temperature decreases from surface to bottom. This processes is reversed

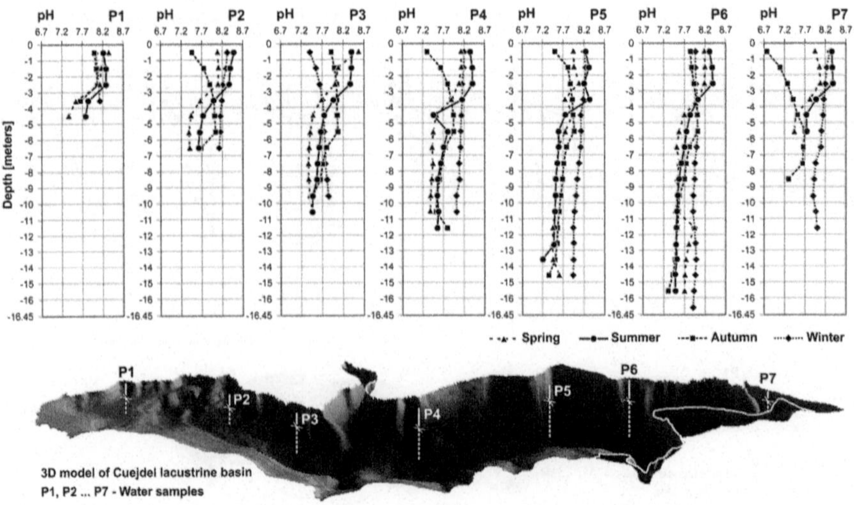

Fig. 7.3 Seasonal distributions of vertical pH values (pH units) in the Cuejdel lacustrine basin (Mihu-Pintilie et al. 2014a)

during winter. All year round, the temperature from the maximum depth of the lake is 4–5 °C.

At surface, pH value reaches a maximum of 8.6 during spring. The minimum value drops to 6.7 during autumn. Usually the values remain constant between 8.0 and 8.3. The surface minimum value is 6.7. At the lowest depths, values between 7.2 and 8.0 during autumn and winter are recorded. At the surface, the pH experiences a more significant gap compared to the deep waters. At the depth of 2–3 m, a sudden drop of pH is recorded, from 8.5 to 7.5–7.7. The most obvious disturbances are recorded in the dam area, where the vortex is located (Fig. 7.3).

The vertical variation of electrical conductivity produced by the presence of salts increases inversely to the quantity of dissolved oxygen (LDO). The increase tendency of TDS is clear in the case of all the samples collected after the depth of 2–4 m. The total quantity of dissolved salts is lower in the epilimnion, as a result of permanent circulation of water and the highest values are recorded at the hypolimnion. At the water surface the annual average value is 185 mg/L, while at the lacustrine bottom is around 241 mg/L. After the depth of 10 m, where the dissolved oxygen is missing, the water temperature is constant throughout the year (5 °C), and the CDC ranges between 200 and 400 mg/L (Fig. 7.4).

On-site evaluation of physico-chemical parameters of Cuejdel Lake demonstrates the fact that it belongs to the mountain lake category and provides an image of a limnosystem with special features due to its formation and ecological characteristics. The position in the upper part of hydrographic basin, with a specific microclimate, affects the thermal regime. The outlet of lake tributaries and torrential gullies provide a small amount of water, which leads to a slow circulation of water, especially in the lower layer.

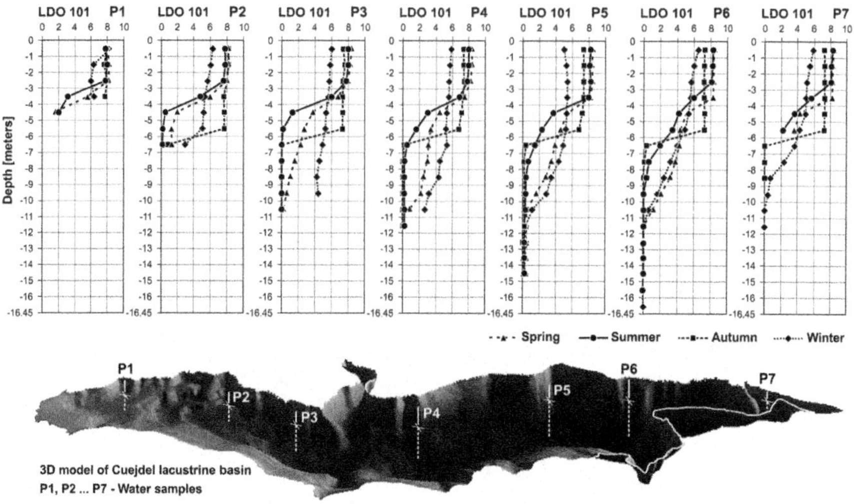

Fig. 7.4 Seasonal distributions of vertical luminiscent dissolved oxygen values (LDO101: mg/L) in the Cuejdel lacustrine basin (Mihu-Pintilie et al. 2014a)

Cuejdel Lake has a slightly alkaline water, and the quantity of dissolved oxygen is specific to mountain rivers, therefore the lake does not have the particularities of an enclosed lake, due to its supplies being provided by the streams, most of them with permanent flow but also due to the water release in the outlet section. Also, the medium degree of water mineralization triggers the presence of a relative balance between the organic component and the mineral suspensions from the lake. At the same time, there is a clear connection of the variations caused by the decrease or increase of depth between temperature, pH, dissolved oxygen and degree of water mineralization during the three seasons (spring, summer, autumn). In winter, the values of these physicoare different due to ice bridge that can last up to 4–5 months. Usually, its thickness is uniform and the tree trunks at surface of the lake supports the ice bridge, especially in the middle section.

A low discharge of Piciorul Crucii and Rotund tributaries leads to higher temperatures on the right bank of the lake. In the area of the Glodu, Cuejdel and Fagetul tributaries, due to a high flow, the water temperatures are relatively equal with the thermal gradient of epilimnion layer of lake. The temperatures during spring and summer have the same tendency and similar values. The water temperature at the surface of lake is generally lower with 2–3 °C than the air temperature. In autumn, the surface temperatures fluctuate between 13 and 14 °C.

The gradual drop of temperature along with the increasing depth is a direct consequence of the connection with the exterior temperature. The existence of relatively uniform temperatures is determined by the thermal homogenization of water. The thermal differences recorded during summer, spring, autumn are small, namely 3–4 °C. It is worth mentioning the phenomenon of water temperature homogenization during autumn. The values are constant, with 13–14 °C at the surface and 11 °C at the depth of 5 m. The homogenization can be created by higher stream flows or the existence of local winds and the disappearance of aquatic vegetation, that lead to an almost perfect homogenization of the water mass.

The thermal homogenization during winter is decreased, being observable on a gap from 1–2 °C to 4–5 °C. Up to the depth of 1 m, temperatures are constant with values of 0–2 °C determined by the ice bridge. At 2 m depth, temperature increases rather rapidly, reaching 2–2.5 °C. Below 3.5 m depth and until the maximum depth of lake, the temperature is constant to 4–5 °C. In spring, summer and autumn, the temperature decreases in direct proportion to the depth. The ice bridge presence prevents the formation of a winter lacustrine basin microclimate. The values for the three seasons, for a depth over 9–10 m are situated around 5 °C (Fig. 7.5).

The optimum concentration of dissolved oxygen is essential for aquatic ecosystems. In spite of this, Cuejdel Lake is not a place thriving with extraordinary biological diversity as its existence is still relatively recent. At the same time, the thermal stratification is correlated with the decrease of oxygen at depth and different life environment appear. High values of dissolved oxygen are recorded in warm seasons while lowest values are present during winter, as a result of the ice bridge. Significant concentrations of dissolved oxygen are caused by streams supplying the aquatorium which create an upstream-downstream current that refreshes the surface waters. The deep circulation is really reduced.

Annual thermal amplitude determines significant differences between cold season and warm season correlated with the increasing of the depth. During summer, at depths of 4 m, the values drop under 4 mg/L. From 6 m, the values reach 2 mg/L. Under 10 m deep, the concentration of dissolved oxygen is <1 mg/L. Considering this aspect, Cuejdel Lake presents an euxinic stratification, with two very different layers, first, at the surface, the one that nurtures life and the deep one, in which the number of life forms drops significantly. In the deepest sections, the most important disturbances of threshold limits take place. The thresholds in this case are extended to values between 4 and 9 m. Between these limits, the differences in value are high. At the depths over 9–10 m the same values are similar for all seasons.

During spring, summer and autumn, it is noticed a direct correlation between LDO, temperature and pH, decreasing from the surface to the depth. These parameters are out of balance during winter, due to the ice bridge that can endure for 15–60 days/year. The long endurance period of the ice bridge is generated by the winter thermal inversions that bring a cold air front on the bottom of the depression. An obvious inconsistency is recorded during winter, when the temperature curve is in reverse (Fig. 7.6).

The quality of aquatic environments is also determined by minerals concentration and organic compounds, particles in suspensions as well as life forms that live in lacustrine biocoenosis. Total dissolved solids (TDS), is a frequently used parameter for identifying the degree of water mineralization and water conductivity. For Cuejdel Lake, the analysis of standard conductivity (CDC401) was made at the same time and in the same points as the other physico-chemical parameters measured.

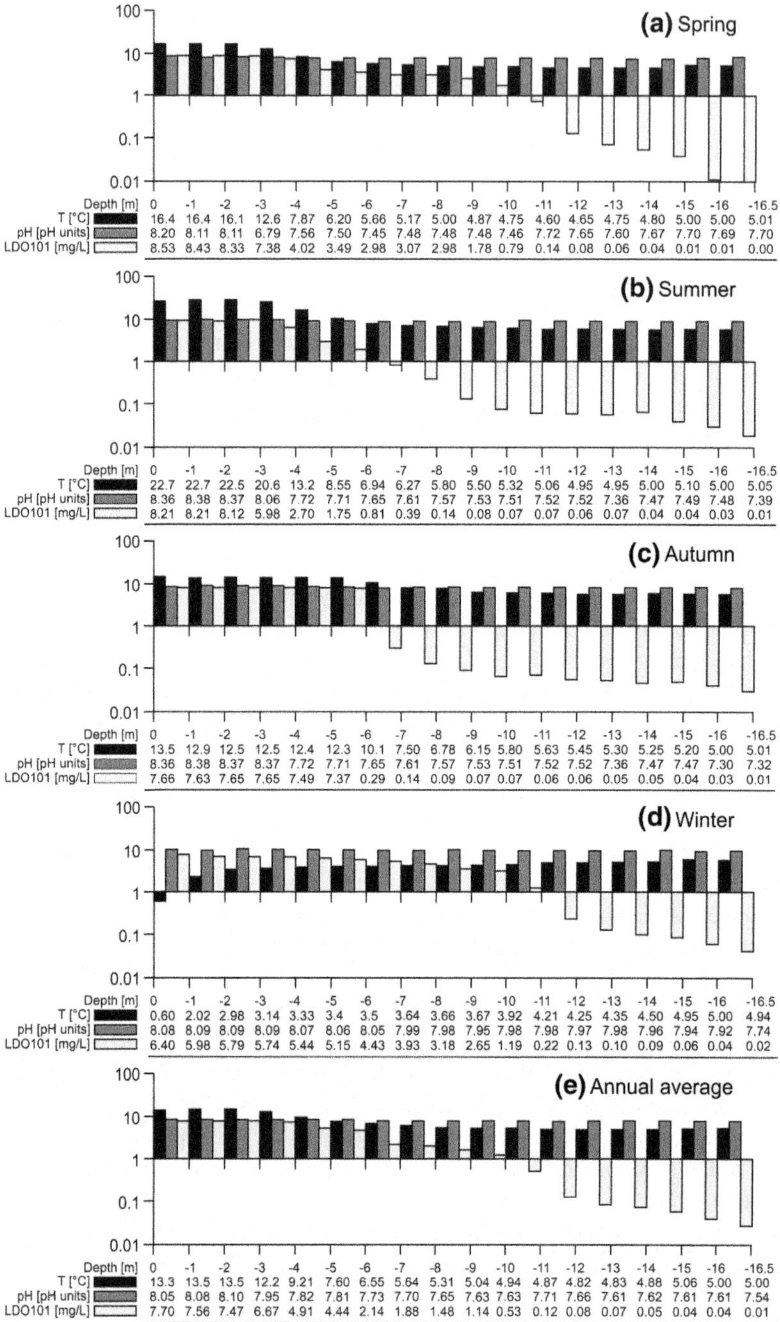

Fig. 7.5 The vertical distribution of temperature (°C), pH (pH units) and luminiscent dissolved oxygen (LDO101: mg/L) during the: **a** Spring; **b** Summer; **c** Autumn; **d** Winter; **e** Annual average (Mihu-Pintilie et al. 2014a)

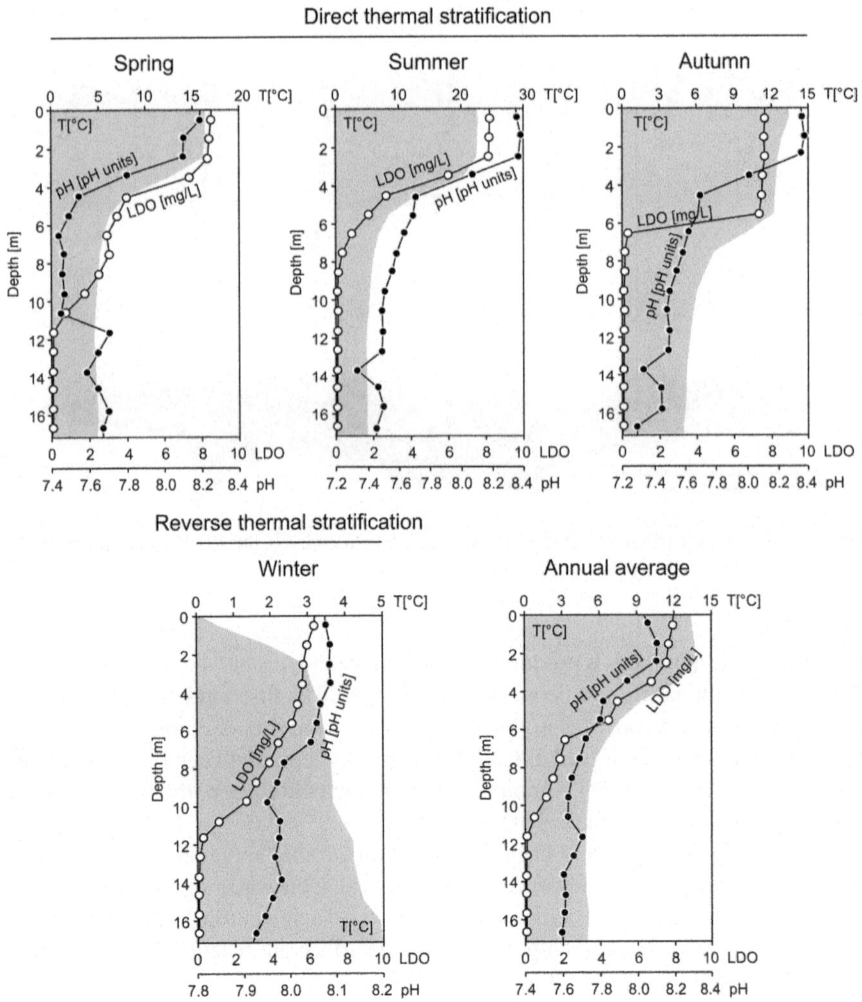

Fig. 7.6 The annual and seasonal variation of the temperature (°C), pH (pH units) and Luminiscent Dissolved Oxygen (LDO101: mg/L) in Cuejdel Lake (Mihu-Pintilie et al. 2014a)

The lake has average values of standard conductivity (CDC401) and implicitly of water mineralization degree (TDS) in normal amounts, between 160 and 220 mg/L all year round. The maximum values are between 270 and 400 mg/L and were recorded for samples P3, P4, P5 and P6, with the highest frequency during autumn, when on the bottom of lacustrine depression the aquatic vegetation begins to decompose. The minimum values are between 160 and 190 mg/L and are specific for the upper part of all samples (Fig. 7.7).

The dissolved oxygen concentration in Cuejdel Lake depends on temperature and water dynamics. At the surface is recorded a maximum of 8.2 mg/L during summer

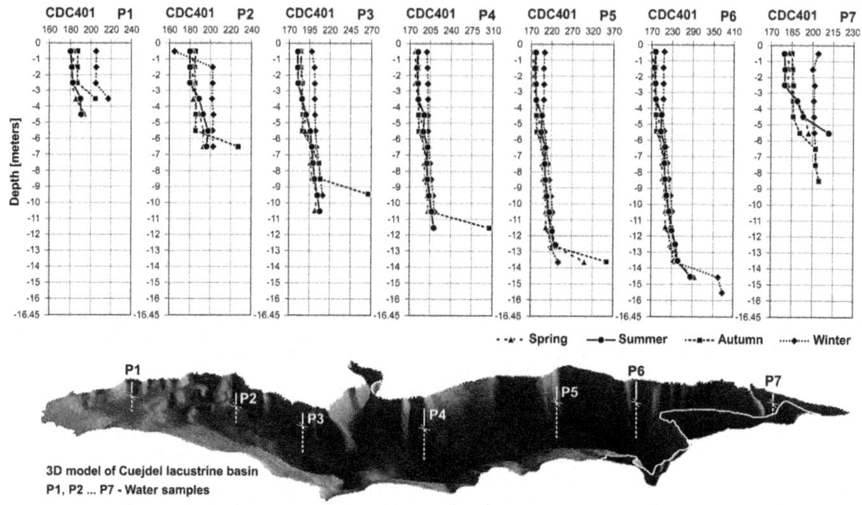

Fig. 7.7 Seasonal distributions of vertical standard conductivity (CDC401: mg/L) values in the Cuejdel lacustrine basin (Mihu-Pintilie et al. 2014a)

and spring. The minimum is produced during winter and reaches 4.5 mg/L. In autumn there are transitional values, between 7 and 8 mg/L. The first sudden change, with a significant drop is recorded 2 m deep. The values from 4 m deep reach 2–4 mg/L. The total absence of dissolved oxygen is felt at 6 m deep. Between 6 and 16 m, the dissolved oxygen is missing completely. During the cold season, the vertical values of physico-chemical parameters are uniform.

Considering the location of Cuejdel Lake within a small mountain valley, thermal inversions are frequently developed. Due to this aspect, the water temperatures cannot be correctly correlated with air temperatures recorded at meteorological stations from Stânişoarei Mts. Maximum temperatures at the surface are reached during summer (22.77 °C) and spring (16.45 °C).

7.2 The Water Quality Index (WQI) and the Trophicity Degree Determined Based on Nutrients

The water quality index (WQI) represents a physico-chemical and biological evaluation of water bodies according to the report between natural state and anthropic influence. Usually, the equation for calculating WQI involves the following parameters: acidifying regime (pH), oxygen regime (DO_2, CBO_5) and nutrients concentration (NH_4^+, NO_3^-, PO_4^{3-}) (Brown et al. 1972).

The method to determine the WQI is based on the equation that provides for each parameter a certain sub-index (Q_i) (Eq. 7.1). The global WQI is calculated through

Table 7.1 Weighting factors for the parameters used in computing WQI (Bharti and Katyal 2011; Mihu-Pintilie et al. 2014c, 2015)

Parameter	Weight (W_1)
Dissolved oxygen (DO_2)	0.17
pH (pH units)	0.11
Biochemical oxygen demand (BDO_5)	0.11
Ammonium ions (NH_4^+)	0.10
Nitrates (NO_3^-)	0.10
Phosphates (PO_4^{3-})	0.10

Table 7.2 WQI classes according to method developed by Kankal et al. (2012)

WQI score	Status
91–100	Excellent water quality
71–90	Good water quality
61–70	Medium high water quality
51–60	Medium low water quality
26–50	Poor water quality
<25	Very poor water quality

calculating each (Q_i) that corresponds to all parameters in the formula, using the weight values from Table 7.1 (Mihăescu et al. 2013; Paiu and Breabăn 2014).

Equation (7.1), The global WQI equation:

$$WQI = \sum_{i=1}^{n} Q_i W_i \tag{7.1}$$

where: Q_i = sub-index calculated for each parameter of water quality; W_i = the weight associated for each parameter of water quality; n = number of parameters used in determining the quality of water.

To synthesize the results achieved a scale from 1 to 100 is used, where 1 is the lowest score and 100 is the highest score. According to studies published in the professional literature, this interval can be divided in many quality classes (Brown et al. 1972; House and Ellis 1987; Bharti and Katyal 2011). Thus, if the global value of WQI is between 91 and 100, the water stream belongs to an excellent quality class and has an unpolluted, natural status. If the values are between 71 and 90, the water quality is good and is affected by a minimum alteration compared to its natural state. The medium quality class (51–70) suggests a moderate alteration, caused by accidental contaminations. A value of WQI situated >50 corresponds to a water body in which frequent dumps of residual materials are made and the chemical parameters of the water are diverted from natural conditions (Kankal et al. 2012) (Table 7.2).

The WQI calculation of Cuejdel Lake was based on 12 water samples collected in polyethylene containers, out of which 7 samples were taken at 1 m deep (epilimnion)

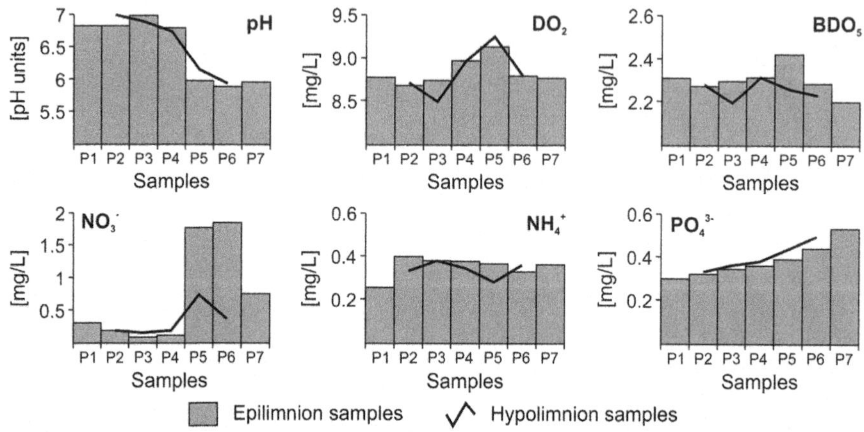

Fig. 7.8 Variation of pH, dissolved oxygen (DO_2), biochemical oxygen demand (BDO_5), nitrates (NO_3^-), ammonium ions (NH_4^-) and phosphates (PO_4^{3-}) on the Cuejdel Lake

and 5 samples at depths between 3 and 16 m (hypolimnion). The sampling took place at the end of warm season to record the optimum period of organic loading of surface water (Fig. 7.8). To validate and compare the results regarding the quality of lacustrine waters, another 30 samples were collected and analysed from downstream the lake, mostly from Cuejdel River until the confluence with Bistrița River (Fig. 7.9).

Tests have been based on in situ determination of physico-chemical parameters: T(°C), pH (pH units), conductivity (mS/cm) and total dissolved solids, using Multi 350i/SET WTW multiparameter instrument, Hanna Hi 4421 bench meter DO/BOD/SOUR, Titroline 700 (SI Analytics) and WTW Turb 555IR, ion-selective electrodes and Shimadzu UV/VIS UV1601 Spectrophotometer for: DO (mg O_2/L), BOD_5 (mg O_2/L), total alkalinity/acidity (mg/L), turbidity (NTU), chlorine (mg/L), total hardness (mg/L), calcium (mg/L) ammonium (mg N/L), nitrate (mg N/L) and phosphates (mg P/L) (Breabăn et al. 2012; Mihu-Pintilie et al. 2014c, 2015).

First parameter entered into the WQI equation was pH as it is an indicator that varies depending on photosynthesis and flora respiration and aquatic fauna. At the moment of sampling, the pH varied between 5.9 and 6.9, with an average of 6.7. The lower values compared to the annual average are caused by respiration and slower decomposing processes in November.

The dissolved oxygen (DO_2) is extremely important for the aquatic fauna respiration, especially for the lacustrine ichthyofauna that needs a concentration of at least 3–5 mg/L to survive. In Cuejdi River, downstream the lake, small variations of DO_2 concentration were recorded, between 13.2 and 20.7 mg/L, while in the Cuejdel Lake the maximum value recorded was 8.83 mg/L. This behaviour is in accordance with the aquatic processes that take place depending on the movement of water or are influenced by the biological activity (Gorde and Jadhav 2013).

The biochemical consume of oxygen (CBO_5) expresses the necessary quantity of DO_2 for the aquatic organisms to decompose the organic matter. Indirectly, the

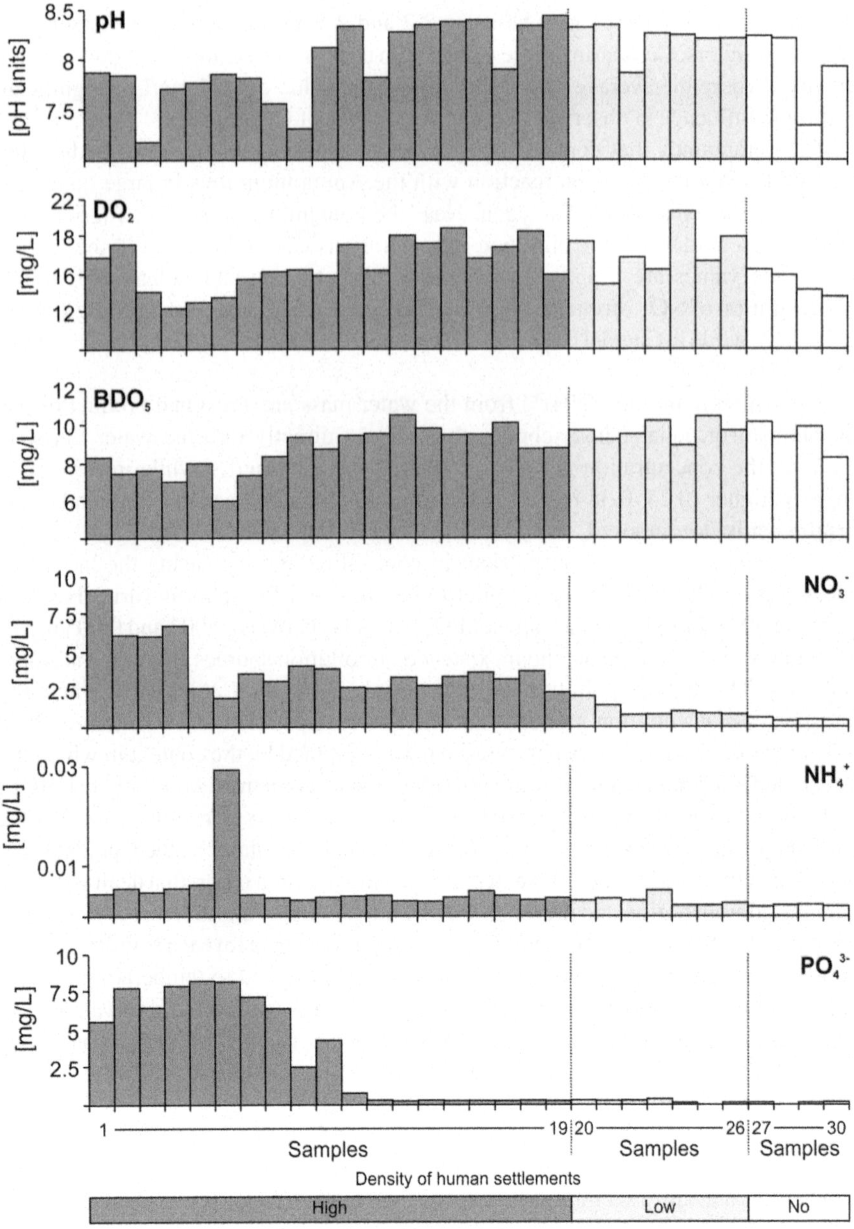

Fig. 7.9 Variation of pH, dissolved oxygen (DO_2), biochemical oxygen demand (BDO_5), nitrates (NO_3^-), Ammonium ions (NH_4^-) and Phosphates (PO_4^{3-}) on the Cuejdi River

CBO_5 indicates also the degree of water loading with organic matter. As the value of biochemical consume of oxygen is lower, the quality of water is better. In the lake, the value of CBO_5 varies between 2.2 and 2.5 mg/L, showing the absence of organic polluters. Comparing these values with the CBO_5 potential from downstream the lake, where the average is situated around the value of 8.94 mg/L, a significant anthropic influence is observed in the water quality of Cuejdi River.

The compounds that contain nitrogen act as nutrients in the water bodies, but the nitrates and the Nitrogen reaction with the Ammonium ions in large quantities determine the exhaustion of oxygen. Near the contamination sources (agricultural lands, household, etc.), the Nitrogen concentrations can be high, but in the natural waters the values are usually lower than 1 mg/L (Gorde and Jadhav 2013). The concentration of NO_3^- from the Cuejdel River varies between 0.4 and 8.9 mg/L (avg. 2.7 mg/L), while in Cuejdel Lake the nitrates concentration is very low, 0.08–1.8 mg/L (avg. 0.5 mg/L).

The Ammonium ions (NH_4^+) from the water mass are a residual product of fish metabolism or aquatic invertebrates that excrete directly into the water mass. In the river the concentration is very low (0.0017–0.0288 mg/L), while in the lake is slightly higher (0.25–0.38 mg/L), due to animal waste and aquatic flora and fauna insufficiently decomposed.

Phosphorus is an essential nutrient in controlling and increasing the lacustrine ecosystem biodiversity. In the Cuejdel Lake, the Total Phosphorus varies between 0.006 and 0.017 mg/L and the value of PO_4^{3-} level is between 0.003 and 0.005 mg/L. Very low values can indicate the inexistence of pollution sources from the anthropic activities. This type of pollution is frequent only along the Cuejdi River where the phosphates concentration gradually increases from upstream to downstream, from 0.02 to 8.1 mg/L. These concentrations can be explained by the context in which the threshold from where the waters are considered to be contaminated is only 0.1 mg/L.

In order to compare and validate the chemical analyses results from 2013 a data range was also used connected to the chemical specificity of the Cuejdel Lake obtained between 2000 and 2006 and published in a research report that dealt with sustainable administration and preservation of Cuejdel Lake Natural Reserve (Research contract CNCSIS 42–2007); therefore, the comparative data reveal a stability of variations for the main physico-chemical parameters connected to temperature (T °C), pH and value of dissolved oxygen (DO_2) between 2000 and 2013. These values suggest an adjustment in time of the lacustrine biotope to the environmental conditions in which the specific processes for aquatic ecosytems were shaped (Mihu-Pintilie et al. 2014a, b, c, 2015).

Regarding the value of nutrients and implicitly of appearances or disappearances of anthropic pollution sources, low values were observed, on a relatively equal variation gap for the entire sampling period (NO_3^- = 0.84–1.22 to 0.08–1.8 mg/L; NO_2^- = 0.013–0.079 mg/L; NH_4^+ = 0.11–1.26 to 0.25–0.38 mg/L; PO_4^{3-} = 0.003–0.004 to 0.003–0.005 mg/L). That is why we consider the chemical parameters indicate a lacustrine environment still untainted by the human society activities and we interpret the value of chemical compounds as a result of a natural contamination in the case of Cuejdel Lake (Table 7.3).

After applying the formula for determining the WQI based on 12 water samples from the lake, a score between 82 and 89 resulted, with an average of 86. The values

Table 7.3 The hydrochemical characteristics of Cuejdel Lake between 2000 and 2014 (Mihu-Pintilie et al. 2014a, b, c, 2015)

Parameter	Unit of measurement	Analysis interval (years)	
		[a]2000–2006	2012–2014
pH	pH units	7.3–7.8	5.9–6.9
Dissolved oxygen (DO_2)	mg/L	4.90–9.04	8.39–8.83
Biochemical oxygen demand (BDO_5)	mg/L	–	2.2–2.5
Organic matter ($KMnO_4$)	mg/L	10.30–19.47	–
Nitrates (NO_3^-)	mg/L	0.84–1.22	0.08–1.8
Nitrites (NO_2^-)	mg/L	0.013–0.079	–
Ammonium ions (NH_4^+)	mg/L	0.11–1.26	0.25–0.38
Phosphates (PO_4^{3-})	mg/L	0.003–0.004	0.003–0.005
Total Phosphorus (TP)	mg/L	0.005–0.015	0.006–0.0017
Standard conductivity (CDC401)	mg/L	–	160–220
Calcium (Ca^{2+})	mg/L	58.5–62.5	–
Magnesium (Mg^{2+})	mg/L	6.8–9.7	–
Bicarbonate (HCO_3^-)	mg/L	201.3–207.4	–
Sulphates (SO_4^{2-})	mg/L	24.3–26.2	–
Chlorine (Cl^-)	mg/L	6.0–7.1	–

[a]After CNCSIS Research Contract No. 42/2007

place Cuejdel Lake in the "Good" category because more than half of analysed samples are close to the lower limit of "Excellent" class. The case of the other 30 samples collected along the Cuejdel stream, downstream from the Cuejdel Lake, the results place the river water into three quality categories: "Good", "Medium-Good" and "Medium-Poor" (Fig. 7.10).

The value distribution shows a degradation of WQI from upstream to downstream, where the minimum values are recorded at the outlet of Cuejdi into Bistrița River (52.58), and the maximum value corresponds to the upper section, close to the confluence of Cuejdel stream with Cuejdi River (78.49). The WQI average for the whole studied section of the river is 62.76, placing the water stream at the lower limit of "Medium-Good" class (Mihu-Pintilie et al. 2015).

In the altitudinal projection of WQI on the hydrographic basin territory of Cuejdi River, the degradation of water quality takes place from upstream to downstream; this type of anthropic pressure is specific for all hydrographic basins from the Carpathian and Subcarpathian areas. In some cases, the contamination of water resources takes

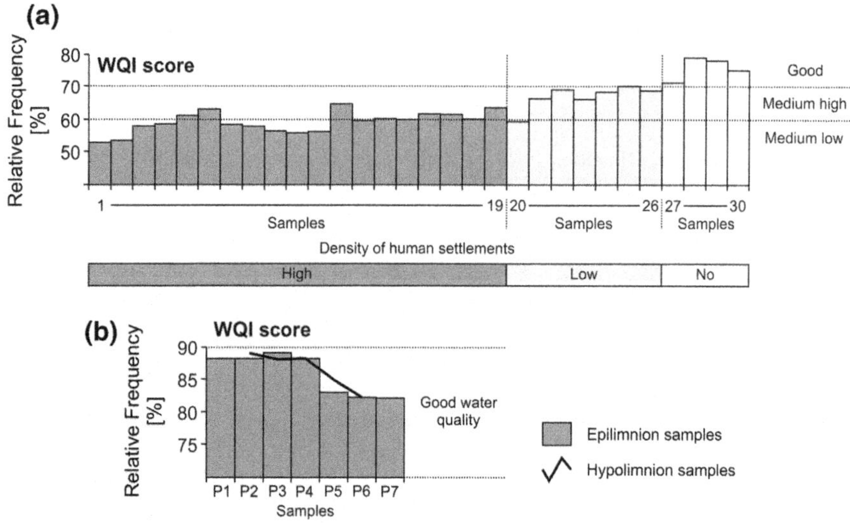

Fig. 7.10 Score of water quality index in: **a** Cuejdi River and **b** Cuejdel Lake

place towards the river springs due to massive deforestation, as a result of forest machinery (hydrocarbon consume) and dumping of wooden waste (swadust, branches, logs). In the case in point, this type of pollution is not made in the Cuejdel Lake area, instead it takes place under the 550 m altitude, on the first terraces of Cuejdi River. Under 450 m, degradation of water quality is linked to the anthropic pressure expressed through dumping of used waters from the agricultural-zootechnical activities (Mihu-Pintilie et al. 2015).

Also, based on nutrients (Total Nitrogen; Total Phosphorus) resulted from the chemical analysis, the trophicity degree of waters can be identified. According to the trophicity degree, the typology of lacustrine units consists of the following main classes: hypertrophic, eutrophic, mezotrophic, oligotrophic and ultraoligotrophic with additional intermediate categories.

Eutrophic lakes hold a primary productivity importance with a high content of nutrients, while the oligothrophic lakes have a low content of nutrients, a consequence of low productivity. Mezotrophic lakes have a medium productivity. Hypertrophic and ultraoligotrophic lakes are specific to aquatic ecosystem evolution that usually tends naturally towards eutrophication (Rogora et al. 2001; Kopácek et al. 2003; Coops et al. 2008; Rzychoń and Worsztynowicz 2008).

Geographical position of lakes within distinct landscape units of Romania (mountain areas, the plateau and plain area, the seaside) create different backgrounds for expressing the trophicity. Concerning the lakes formed in the Eastern Carpathians area, it can be observed that, depending on nutrient values, most of them belong to mezotrophic, mezo-eutrophic, eutrophic, eutro-hypertrophic and hypertrophic categories (>85%). Cuejdel Lake is an mezotrophic aquatorium, with similar trophicity

Table 7.4 Morphometric characteristics and trophicity based on nutrients (total nitrogen and phosphorus) of some natural lakes in Eastern Carpathians (Romanescu et al. 2013; Mihu-Pintilie et al. 2014b)

Name	Lake type	River/stream	Alt. (m)	Vol. (m^3)	Area (ha)	Depth (m)	Trophicity base on nutrients
Buhăescu	Glacial	Tisa	1905	4000	0.2	5	Mezo–Eutrophic
Lala Mare	Glacial	Someş	1815	–	0.56	2	Ultraoligotrophic
Bodi	Flood plain	Săsar	731	300,000	1.6	7	Eutro–Hypertrophic
Saint Ana	Volcanic	Olt	946	250,000	19.5	7.5	Mezotrophic
Red Lake	Natural dam	Bicăjel	965	720,000	12.05	10.5	Mezo–Eutrophic
Cuejdel	Natural dam	Cuejdel	661	920,000	13.95	16.5	Mezotrophic
Constellation	Natural dam	Cuejdel	615	7000	13.95	4.5	Mezotrophic
Ponoare	Natural dam	Cuejdiu	850	6000	0.34	7.8	Mezo–Eutrophic
Blue Lake	Natural dam	Cuejdel	585	–	0.1	4	Mezotrophic

conditions with those from vulcanic lake Sfânta Ana, or with the anthropic accumulations Brădișor and Izvorul Muntelui (Eastern part of Carpathians).

In conclusion, the position of Cuejdel Lake in the medium altitude mountains, the high degree of forestation of the catchment area and the reduced anthropic influence determine a good, almost excellent quality of water (WQI). The degree of contamination with Nitrogen and Phosphorus compounds (nitrites, nitrates, phosphates, etc.) is very low and the value of nutrients reveal a natural functioning of the limnosystem. Also, due to the recent formation of the lake (1991) and the low level of lacustrine biocoenosis, the aquatic ecosystem belongs to the medium class of trophicity. Still, the degradation of its functionality is derived from the fact that nowadays an increase in forest exploitation is ongoing in the catchment area of the lake; a serious warning should be raised concerning the destructive potential of these activities on the lacustrine biodiversity through water degradation. In the same context, extending the deforestated parcel could lead to acceleration of erosion processes that irrecoverably degradates the newly created natural habitat (Table 7.4).

References

Aplay S, Veillette JJ, Dixit AS, Dixit SS (2006) Regional and historical distributions of lake-matter pH within a 100-km radius of the Horne smelter in Rouyn-Noranda, Quebec, Canada. Geochem Explor Environ Analys 6(2–3):179–186

Araoye PA (2009) The seasonal variation of pH and dissolved oxygen (DO2) concentration in Asa lake Ilorin, Nigeria. Int J Physic Sci 4(5):271–274

Belevantser VI, Ryzhikh AP, Smolyakov BS (2008) Diurnal and vertical variability of pH (O_2), and Eh in the Novosibirsk water reservoir. Russian Geol Geophys 49:673–681

Bharti N, Katyal D (2011) Water quality indices used for surface water vulnerability assessment. Int J Environ Sci 2(1):154–173

Breabăn IG, Gheţeu D, Paiu M (2012) Determination of water quality index of Jijia and Miletin Ponds. Bull UASVM Agric 69(2):160–167

Brown RM, McLelland NI, Deininger RA, O'Connor MF (1972) A water quality index—crashing the psychological barrier. Ind Environ Quality 1:173–182

Bukreev VI (2013) Effect of the non-monotonic dependence of the water density on temperature on convection induced by cooling from above. Aquat Sci Technol 1(1):52–65

Chapman LJ, Chapman CA, Srisman TL, Nordlie FG (1998) Dissolved oxygen and thermal regimes of a Ugandan crater lake. Hydrobiologia 385:201–211

Coops H, Buijse L, Buijse ADT, Constantinescu A, Covaliov S, Hanganu J, Ibelings BW, Menting F, Năvodaru I, Oosterberg W, Staras M, Torok L (2008) Trophic gradients in a large river Delta: ecological structure determined by connectivity gradients in the Danube Delta (Romania). River Res Appl 24(5):698–709

Dake JMK, Harleman DRF (1969) Thermal stratification in lakes: analytical and laboratory studies. Water Resour Res 5:484–495

Davis RB, Anderson DS, Whiting MC, Smol JP, Dixit SS (1990) Alkalinity and pH of Three Lakes in Northern New England, USA, over the past 300 year. Philosoph Trans Royal Soc Bull 327(1240):413–421

Edingher JE (1970) Vertical temperature structure and water surface heat exchange. Water Resour Res 6:1392–1395

Enea A, Stoleriu CC, Romanescu G (2011) GIS techniques used in the analysis of the morphometrical parameters of the Red Lake drainage basin. Luc Semin Geogr Dimitrie Cantemir 31:39–48

Enea A, Romanescu G, Stoleriu C (2012) Quantitative considerations concerning the source-areas for the silting of the Red Lake (Romania) lacustrine basin. In: Gâştescu P, Breţcan P (eds) Proceedings conference of the water resources and wetlands, Tulcea, 11–13 September 2012, pp 119–123

Gâştescu P (1971) Lacurile din România–limnologie regională. Romanian Academy Publishing House, Bucharest

Genova SN, Belolipetskii VM, Rogozin DY, Degermendzhy AG, Mooij WM (2010) A one-dimensional model of vertical stratification of Lake Shiva focused on water conditions and ice cover. Aquat Ecol 44:571–584

Glazik R, Marszelewski W, Skowron R (2006) Selected problems of registration of vertical distribution of water temperatures in lakes. Limnological Rev 6:103–110

Gorde SP, Jadhav MV (2013) Assessment of water quality parameters: a review. J Eng Res Appl 3(6):2029–2035

House MA, Ellis JB (1987) The development of water quality indices for operational management. Water Sci Technol 19:45–154

Kankal NC, Indurkar MM, Gudadhe SK, Wate SR (2012) water quality index of surface water bodies of Gujarat, India. Asian J Explor Sci 26(1):39–48

Kopácek J, Cosby BJ, Majer V, Stuchlík E, Veselý J (2003) Modelling reversibility of central European mountain lakes from acidification: 2nd Part—the Tatra Mountains. Hydrol Earth Syst Sci 7(4):510–524

Lampert W, Sommer U (2007) Limnoecology—the ecology of lakes and streams, 2nd edn. Oxford University Press, New York

Maberly SC (1996) Diel, episodic and seasonal changes in pH and concentration of inorganic carbon in a productive lake. Frashwater Biol 35(3):579–598

Malm J, Zilitinkevich S (1994) Temperature distribution and current system in a connectively mixed lake. Bound Layer Meteorol 71(3):219–234

Maslanka W (2009) Atypical summer vertical diversity of physical-chemical properties of lake waters. Limnological Rev 9(2–3):91–96

Mihăiescu T, Mihăiescu R, Vârban D, Vârban R, Mihăiescu M (2013) Water quality assessment of the Nadas river (Romania) in terms of NSF water quality index. Ann Oradea Univ-Fascicle Enviro Protect 21:651–656

Mihu–Pintilie A, Romanescu G (2011) Morphometric and morphological suitability of the relief from the Crucii (Cuejdel) Lake basin (Stânişoarei Mountains). In: Pandi G, Moldovan F (eds) Proceedings of the air and water components of the environment, Babeș–Bolyai University, Cluj, 18–19 March 2011, vol 1, pp 305–313

Mihu-Pintilie A, Romanescu G, Stoleriu C (2012) Morpho–bathymetric parameters of recess Crucii (Cuejdel) Lake (Stânişoarei Mountains). In: Pandi G, Moldovan F (eds) Proceedings of the air and water components of the environment, Babeș–Bolyai University, Cluj, 23–24 March 2012, vol 1, pp 445–452

Mihu-Pintilie A, Romanescu G, Stoleriu C (2014a) The seasonal changes of the temperature, pH and dissolved oxygen in the Cuejdel Lake, Romania. Carpat J Earth Enviro Sci 9(2):113–123

Mihu-Pintilie A, Paiu M, Breabăn IG, Romanescu G (2014b) Status of water quality in Cuejdi hydrographic basin from Eastern Carpathian, Romania. In: SGEM—Proceedings of the 14th international multidisciplinary scientific geoconferences—hydrology and water resources, 18–24 June, Albena, vol 14, no 1, pp 639–646

Mihu-Pintilie A, Romanescu G, Stoleriu C, Stoleriu O (2014c) Ecological features and conservation proposal for the largest natural dam lake in the Romanian Carpathians-Cuejdel Lake. Int J Conserv Sci 5(2):243–252

Mihu-Pintilie A, Romanescu G, Stoleriu CC, Breaban IC (2015) Physico-Chemical Parameters in Mountain Freshwater: Cuejdiu River from Eastern Carpathians, Romania. Key Eng Mater 660:257–261

Moalla SMN (1996) Seasonal variation of high dam lake water. J Environ Sci Health-Part A 31(4):731–746

Moundiotiya C, Sisodia R, Kulshreshthg M, Bhatra AL (2004) A case study of the Jawwa Ramgarh wetland with special reference to physico-chemical properties of water and its environs. J Environ Hydrol 12(24):1–7
Newcombe CL, Dweyr PS (1949) An analysis of the vertical distribution of temperature in a dichothermic lake of southern Michigan. Ecology 30(4):443–449
Nolan M, Brigham-Grettej J (2007) Basic hydrology, limnology, and meteorology of modern Lake El'gygytgyn, Siberia. J Paleolimnol 37(1):17–35
Paiu M, Breabăn IG (2014) Water quality index—an instrument for water resources management. In: Pandi G, Moldovan F (eds) Proceedings of the air and water components of the environment, Babeș–Bolyai University, Cluj, 23–24 March 2014, vol 1, pp 391–398
Pandi G, Magyari ZS (2004) Realizarea hărților batimetrice pe calculator. Modelul Lacul Roșu. Stud Univ Babeș-Bolyai Cluj 1:55–60
Pandi G, Buzilă L (2004) Caracteristici hidro–geomorfologice ale sedimentarii in Lacul Rosu. In: Geography within the Context of Contemporary Development, Cluj-Napoca University Press, Cluj
Patra AP, Patra JK, Muhapatra NK, Das S, Awain GC (2010) Seasonal variation in physicochemical parameters of Chilika Lake after opening of new mouth near Gabakunda, Orissa. India. World J Fish Marine Sci 2(2):109–117
Ravens TM, Kocsis O, Wüest A (2000) Small-scale turbulence and vertical mixing in Lake Baikal. Limnol Oceanogr 45(1):159–173
Rădoane N (2002) Un nou lac de baraj natural în bazinul Bistriţei Moldoveneşti-Lacul Cuejdel. Stud Cerc Geograph 49:76–82
Rogora M, Marchetto A, Mosello R (2001) Trends in the chemistry of atmospheric deposition and surface waters in the Lake Maggiore catchment. Hydrol Earth Syst Sci 5(3):379–390
Romanescu G (2009a) Trophicity of lacustrine wetlands on the Carpathian territory of Romania. A case study from the East Carpathian mountains. Luc Sem Geogr D Cantemir, Univ Alexandru Ioan Cuza Iași 29:5–13
Romanescu G (2009b) Trophicity of lacustrine waters (lacustrine wetlands) on the territory of Romania. Lakes–Reservoirs Ponds—Rom J Limn 3:62–72
Romanescu G (2009c) The physical and chemical characteristics of the lake wetlands in the central group of the east Carpathian Mountains. Lakes, Reservoirs Ponds—Rom J Limn 4:94–108
Romanescu G, Stoleriu CC (2010) Parametrii morfobatimetrici ai cuvetei lacustre Roşu (Hăghimaş). In: Gâştescu P, Breţcan P (eds) Lucrările Simpozionului Naţional–Resursele de apă din România, Târgovişte, 11–13 June 2010, vol 1, pp 308–314
Romanescu G, Stoleriu CC (2014) The seasonal variation of temperature, pH and dissolved oxygen concentration in Lake Rosu, Romania. CLEAN—Soil, Air Water 42(3):236–242
Romanescu G, Stoleriu CC, Lupascu A (2012) Biochemistry of wetlands in barrage Lacul Rosu catchment (Haghimas–Eastern Carpathian). Environ Eng Manage J 11(9):1627–1637
Romanescu G, Târnovan A, Sandu I, Cojoc GM, Breabăn IG, Mihu-Pintilie A (2015) Water chemism within the settling pond of Valea Straja and the quality of the Suha water body (Eastern Carpathians). Rev Chim (Bucharest) 66(10):1700–1706
Rucinski DK, Beletsky D, DePinto JV, Schwab DJ, Scavia D (2010) A simple 1-dimensional, climate based dissolved oxygen model for the central basin of Lake Erie. J Great Lakes Res 26:465–476
Rzychoń D, Worsztynowicz A (2008) What affects the nitrogen retention in Tatra Mountains lakes' catchments in Poland? Hydrol Earth Syst Sci 12(1):415–424
Skowron R (2008) Water thermal conditions during winter stagnation in the selected lakes in Poland. Limnological Rev 8(3):119–128
Sundaram TR, Rehm RG (1972) The seasonal stratification cycle of temperature lakes including the effects of power plant thermal discharge. AIAA J 10:204–210
Stauffer RE (1992) Efficient estimation of temperature distribution, heat storage, thermocline migration and vertical eddy conductivities in stratified lakes. Freshwat Biol 27(3):307–326

Stoleriu CC, Stoleriu O, Mihu-Pintilie A (2014) Scientific and tourist value of natural dam lakes in the Carpathian Mountains (Romania). Case study: Red, Cuejdel and Iezerul Sadovei Lakes. In: SGEM–proceedings of the 14th international multidisciplinary scientific geoconferences–ecology and environmental protection, 18–24 June, Albena vol 14, no 2, pp 625–632

Sweers HE (1968) Two methods of describing the “Average” vertical temperature distribution of a lake. J Fisheries Res Board Canada 25:911–1923

Tadesse I, Green FB, Puhakka JA (2004) Seasonal and diurnal variations of temperature, pH and dissolved oxygen in advanced integrated wastewater pond system treating tonnery effluent. Water Res 38(3):645–654

Thakur SS, Bais VS (2006) Seasonal variation of temperature, alkalinity and dissolved oxygen in the Sagar Lake. Acta Hydrochim Hydrobiol 15(2):143–147

Wand U, Schwarz G, Brüggemann E, Bräuer K (1997) Evidence for physical and chemical stratification in Lake Untersee (central Dronning Maud Land, East Antarctica). Antarctic Sci 9(1):43–45

Wassenaar LI (2012) Dissolved oxygen status of Lake Winnipeg: Spatio-temporal and isotopic ($\delta_{18}O$-O_2) patterns. J Great Lakes Res 38(3):123–132

Xantus I, Xantus J (1999) Hagymas-hegyseg es a Gyilkos-to kornyeke. Pallas Akademia, Csikszereda

Chapter 8
Vegetation and Fauna. Environmental Bioindicators from the Lacustrine Water Body

Abstract The transition from a simple river sector specific to the Eastern Carpathians to a lake basin led to an increase of biodiversity. The assessment of vegetation and fauna from Cuejdel Lake is based on interdisciplinary research in collaboration with biologists. Also, some biological data have been taken from the scientific literature. The results of research led to the determination of 11 habitats, 38 bird species, 8 amphibian species and 2 reptile species in the Cuejdel Lake Nature Reserve. Out of these, in the 2006 modified version (2006/105/CE) of the Habitats (92/43/EEC) and Birds Directives (79/409/EEC) are there is only mention of 5 habitats, 8 species of birds, 3 of amphibians and a single reptile species. The trophicity level in the aquatic environment was based on ecological bio-indices, such as: phytoplankton (50 taxons), vertebrate fauna (17 species) and invertebrate fauna (over 150 taxons). Within the category of invertebrate fauna are included insects (over 100 taxons), zooplankton (14 taxons) and benthic macroinvertebrates (7 taxons). 72 taxons with high density indicate a saprobic value based on dominant or specific valence. Also, have been identified 11 aquatic and palustrine phitocenosis that represent propitious habitats both for the existing species and the new ones. Based on ecological data collected, Cuejdel Lake acts as an oligotrophic aquatic environment with mesotrophic tendencies.

This study also includes a series of research and investigation methods specific for limnobiological studies. The investigation of the biotic environment aimed to depict the lacustrine basin based on ecological bioindicators (macrophytes, ichthyofauna, avifauna, phytoplankton, zooplankton, zoobentos, etc.). A Nansen bottle was used to collect the water samples, and for determining the biological loading of the water a net was used to filter 10 L water/sample. The alluvial samples from the benthic zone were extracted with a mechanical device fitted with preservation tubes through the vacuum method (Valframe—volume sample 0.5 L), and for collecting the samples from the first layer a benthic instrument was used (Van Veen grabs—volume sample 2.5 L). The collected samples were later analysed through specific methods in the laboratory.

The analysis of samples was realized in collaboration with specialists from domains connected to limnology (biologists, chemists, geologists, etc.), active in

A. Mihu-Pintilie, *Natural Dam Lake Cuejdel in the Stânişoarei Mountains, Eastern Carpathians*, https://doi.org/10.1007/978-3-319-77213-4_8

Table 8.1 The summary of laboratory activities to identify the ecological bioindicators from the Cuejdel lacustrine water body (Mihu-Pintilie et al. 2014)

Sample type	Sampling device	Laboratory equipment	Analysis method	Monitoring frequency [samples data/year]
Lacustrine sediments from benthic zone	Van Veen grabs—[vol. 0.5 L]	Microscope	Determination of zoobentos	2/year [2012, 2013]
	Valframe—frame with sample tube guide [vol. 2.5 L]	Microscope	Determination of zoobentos	2/year [2013]
Water from Epilimnion zone of lake	Nansen bottle 10 * [vol. 1 L]	Net/Microscope	Determination of phytoplankton	2/year [2012–2014]
	Nansen bottle 10 * [vol. 1 L]	Net/Microscope	Determination of zooplankton	2/year [2012–2014]

many scientific research departments of Alexandru Ioan Cuza University from Iași. The summary of laboratory activities are depicted in Table 8.1.

8.1 Vegetation, Aquatic and Palustrine Phytocoenoses and Phytoplankton

Vegetation. The Cuejdel Lake was formed in a heavily forested area, where species from the Eastern Carpathians dominate, such as fir (*Abies alba*), spruce (*Picea abies*) and beech (*Fagus sylvatica*), placed in mixture clusters or compact groups. In deforested areas and in areas where trees were knocked down by the wind, thickets of lungwort (*Pulmonaria rubra)* and raspberry tree (*Rubus idaeus*) are representative. The zonation of forest vegetation is atypical, as the beech and spruce reach the same altitude level. This phenomenon is obvious on the left slope of the lake, where the beech and spruce bands are transversally arranged on the lacustrine unit axis and sometimes they alternate and mix. Also, the trees caught in the water mass are usually spruce individuals that dominate the right section of the lake and beech in mixture with specific grove species—their trunks still emerging to the surface.

Between parcels with compact forests and the water body, a tree belt was developed specifically for wet zones, especially on the right bank and upstream the lake where the following species can be observed: white alder (*Alnus incana*), mountain sycamore (*Acer pseudoplatanus*), shaking poplar (*Populus tremula*), osier (*Salix fragilis*), goat willow (*Salix caprea*), hornbeam (*Carpinus betulus*), hazelnut tree (*Corylus avellana*) and ash (*Fraxinus excelsior*). In the dam area and on its surface there are also the elder tree (*Sambucus nigra*) and black alder (*Alnus glutinosa*). The

Fig. 8.1 Distribution of wetlands around Cuejdel Lake (drone capture—September 2016)

outlets into the lake of Cuejdel, Glodu, Piciorul Crucii, Pârâul Rotund and Făgetu creeks maintain the highest biodiversity. On the first terrace of lacustrine abrasion a small meadow evolved, formed from a phytocoenoses with specific species of grasses and vegetables (Fig. 8.1).

The summary of cormophyte elements in the Cuejdel Lake area indicates the following families and species:

- *Aceraceae: Acer platanoides, A. pseudoplatanus*
- *Alismataceae: Alisma lanceolatum, A. plantago-aquatica*
- *Anthyriaceae: Athyrium filix- femina*
- *Aspidiaceae: Dryopteris filix*
- *Balsaminaceae: Impatiens noli-tangere*
- *Betulaceae: Alnus incana*
- *Boraginaceae: Echium vulgare, Myosotis caespitosa, Pulmonaria officinalis, P. rubra, Symphytum cordatum*
- *Callitrichaceae: Callitriche palustris*
- *Caryophyllaceae: Cerastium holosteoides, Stellaria aquatica*
- *Coryllaceae: Carpinus betulus, Corylus avellana*
- *Cruciferae: Alliaria officinalis, Cardamine amara, C. impatiens, Lepidium campestre, Rorippa sylvestris*

- *Euphorbiaceae: Euphorbia amygdaloides, Mercurialis perennis*
- *Equisetaceae: Equisetum arvense, E. Fluviatile, E. telmateia*
- *Fagaceae: Fagus sylvatica*
- *Gentianaceae: Gentiana asclepiadea*
- *Grossulariaceae*: *Ribes rubrum, R. uva- crispa*
- *Hypericaceae: Hypericum hirsutum, H. Montanum, H. perforatum*
- *Lamiaceae: Calamintha clinopodium, Glecoma hederacea, Lycopus europaeus, Mentha longifolia, Origanum vulgare, Prunella vulgaris, Stachys sylvatica*
- *Leguminosae: Astragalus glycyphyllos, Coronilla varia, Lotus corniculatus, Medicago lupulina, Trifolium medium, T. pratense, T. repens, Vicia sepium*
- *Liliaceae: Paris quadrifolia*
- *Lythraceae: Lythrum salicaria*
- *Oleaceae: Fraxinus excelsior*
- *Onagraceae: Epilobium palustre, E. parviflorum, E. roseum*
- *Orchidaceae: Epipactis helleborine*
- *Pinaceae: Abies alba, Picea abies*
- *Polygonaceae: Polygonum hydropiper, Rumex conglomeratus, R. crispus, R. sanguineus*
- *Potamogetonaceae: Potamogeton pectinatus*
- *Primulaceae: Lysimachia nummularia, Primula veris* ssp. *Veris*
- *Pyrolaceae: Pyrola minor*
- *Ranunculaceae: Caltha palustris* ssp. *Laeta, Anemone nemorosa, Hepatica transsilvanica, Ranunculus acris, R. carpaticus*
- *Rosaceae: Cotoneaster integerrimus, Potentilla reptans, Rosa canina, Rubus hirtus, R. idaeus, Sorbus aucuparia*
- *Rubiaceae: Galium odoratum, G. palustre, G. schultesii*
- *Salicaceae: Salix caprea, S. cinerea, S. fragilis, Populus tremula*
- *Umbelliferae: Carum carvi, Torilis arvensis*
- *Urticaceae: Urtica dioica*
- *Violaceae*: *Viola hirta*

This floristic diversity completes the cormophytes species established in the Management Plan of the Cuejdel Lake Protected Area (v1/15.II.2012), based on surveyings that describe the abundance, dominance and participation of species within the phytocoenoses around the lake.

The bioforms spectrum indicates the following spread: hemycriptophytes–46.0%, megafanerophytes–5.8%, microphenarophytes–3.7%, megaphanerophytes/microphanerophytes–2.2%, nanophanerophytes–2.9%, camephytes–2.2%, hemycriptophytes/geophyte–3.7%, geophyte–7.3%, geophyte/helohidatophyte–4.5%, helohidatophyte–5.8%, helohidatophyte/geophytes–1.4%, annual terophytes–5.8%, annual/biannual terophytes–2.9%, biannual terophytes–5.8% (Sanda et al. 2008; Bodescu et al. 2010) (Table 8.2; Fig. 8.2).

According to the research report on biodiversity of the natural dam lake Cuejdel, realized at the Biological Station—Petre Jitaru from Piatra Neamț in 2006,

Table 8.2 The abundance/dominance of species within phytocoenoses around the Cuejdel lacustrine basin (Bodescu et al. 2010; Mihu-Pintilie et al. 2014)

Species	No. of surveys (abundance/dominance)								
	1	2	3	4	5	6	7	8	9	10
Abies alba	+	·	·	·	·	·	·	+	·	·
Actea spicata	·	·	·	·	·	·	·	·	·	+
Aegopodium podagraria	·	+	·	+	·	·	·	·	·	·
Alliaria petiolata	·	+	·	·	·	·	·	·	·	·
Allium ursinum	+	4	·	·	+	·	·	·	·	·
Angelica sylvestris	·	·	·	+	·	·	·	·	·	·
Anemone nemorosa	·	+	·	·	+	+	+	·	·	·
Anemone ranunculoides	+	·	·	·	·	+	·	·	+	
Caltha palustris	+	·	+	·	·	·	·	·	·	+
Cardamine sp.	+	·		+	·	·	·	·	·	·
Carex sp.	·	·	3	·	·	·	·	·	·	·
Carex praecox	·	·	·	·	·	·	+	·	·	·
Chrysosplenium alternifolium	3	·	2	·	·	·	·	·	·	·
Corydalis solida	·	·	·	·	·	·	·	·	·	+
Cruciata glabra	·	·	·	·	·	·	+	·	·	·
Dactyloriza sambucina	·	·	·	·	·	·	+	·	·	·
Daphne mezereum	·	·	·	·	·	·	·	·	·	+
Dentaria glandulosa	·	1	·	·	+	2	·	1	·	·
Dentaria bulbifera	+	·	·	·	+	1	·	1	2	·
Dryopteris filix-mas	·	·	·	·	·	+	·	+	+	·
Euphorbia amygdaloides	·	·	·	·	+	·	·	+	·	·
Fagus sylvatica	+	·	·	·	·	·	+	1	·	·
Fragaria vesca	·	·	·	·	·	+	·	·	·	·
Gagea lutea	·	·	·	·	·	·	·	·	·	+
Galanthus nivalis	·	·	·	·	·	·	·	·	·	+
Galium odoratum	·	·	·	·	+	+	·	1	·	·
Galium aparine	·	·	·	+	·	·	·	·	·	·
Geranium robertianum	·	+	·	·	·	·	·	·	·	·
Geum urbanum	·	+	·	·	·	·	·	·	·	·
Glechoma hirsuta	·	·	·	+	+	·	·	·	·	·
Hepatica nobilis	·	·	·	·	·	·	·	·	·	+
Hepatica transsilvanica	·	·	·	·	·	·	·	·	·	+
Isopyrum thalictriodes	·	+	·	·	+	2	·	·	+	·
Lathrea squamaria	·	·	·	·	·	·	·	·	·	+
Lemna sp.	+	·	·	·	·	·	·	·	·	·

(continued)

Table 8.2 (continued)

Species	No. of surveys (abundance/dominance)									
	1	2	3	4	5	6	7	8	9	10
Lunaria rediviva	·	·	·	·	·	·	·	+	·	·
Lysimachia nummularia	·	·	·	+	·	·	·	·	·	·
Mercurialis perennis	+	·	·	·	·	+	·	+	+	·
Oxalis acetosella	·	·	·	·	+	·	·	1	·	·
Paris quadrifolia	·	·	·	·	+	+	·	·	·	·
Petasites albus	·	·	+	·	·	·	·	·	·	·
Petasites hybridus	·	·	+	·	·	·	·	·	·	·
Primula veris	·	·	·	·	·	·	+	·	·	+
Pulmonaria sp.	+	·	·	·	·	·	·	·	·	·
Pulmonaria rubra	·	·	·	·	·	·	·	·	·	+
Ranunculus acris	·	·	·	+	+	·	·	·	·	·
Ranunculus cassubicus	·	·	·	·	+	·	·	·	·	·
Ranunculus ficaria	+	+	·	·	+	·	·	·	·	·
Ribes uva-crispa	·	·	·	·	·	·	·	·	·	+
Rumex sp.	+	·	·	·	·	·	·	·	·	·
Rubus caesius	+	·	·	+	·	·	·	·	·	·
Salvia glutinosa	·	·	·	+	·	·	·	·	·	·
Scilla bifolia	·	·	·	·	+	·	·	·	·	·
Stellaria sp.	·	·	·	+	·	·	·	·	·	·
Symphytum cordatum	·	·	·	·	1	·	·	·	+	·
Taraxacum officinale	·	·	·	+	+	·	·	·	·	·
Trollius europaeus	·	·	·	·	·	·	+	·	·	·
Tussilago farfara	·	·	·	+	·	·	·	·	·	·
Urtica dioica	+	+	·	·	·	·	·	·	·	·
Viola sp.	·	·	·	·	+	·	+	·	·	·

the floristic elements show a predominance of phytogeographic origine as follows: Euroasian–47.2%; European–14.5%; Central European–5.0%; Circumpolar–13.0%; cosmopolitan–13.0%; adventive–2.2%; Atlantic/Mediterranean–0.7%; Carpato-Balkanic–2.2% and the rest are endemic species (Fig. 8.3).

According to the humidity affinity, more than half of floristic elements are adapted to a wet environment as follows: mezophytes–42.0%, mezohydrophytes–21.7%, xeromezophytes–14.5%, hidrophytes–11.6%, ultrahydrophytes–5.8%, xerophytes–1.5 and 2.9% of the total are amphitolerant to humidity. The affinity to temperature showed the predominance of the following elements: mezoterms–60.0%, microterms–13.8%, moderately thermophile–8.8%, criophile–0.7%, and a density of species regardless of the high density of species that are indifferent to this factor–36.2%. Based on soil affinity, 31.2% of species are adapted

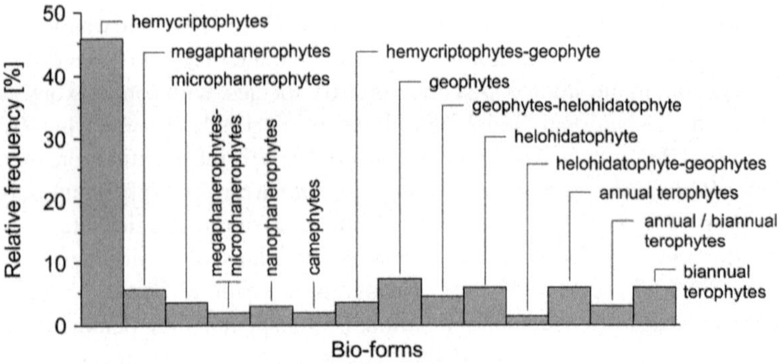

Fig. 8.2 Histogram of the bio-forms distribution within phytocoenosis around Cuejdel Lake

Fig. 8.3 Histogram of the phytogeographic origins of the floristic elements around Cuejdel Lake

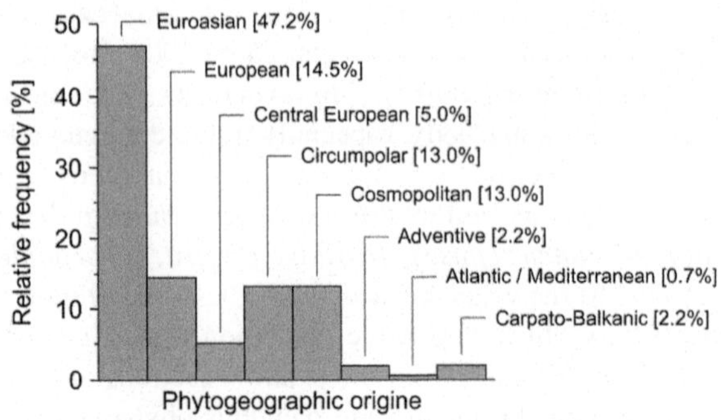

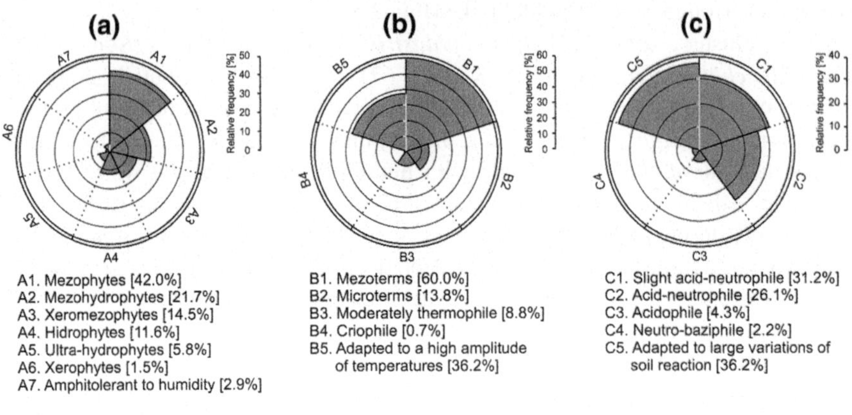

Fig. 8.4 The distribution of floristic elements around Cuejdel Lake depending on: **a** humidity, **b** temperature and **c** soil affinities

to a slight acid-neutrophile reaction, 26.1% of species to a acid-neutrophile reaction, 4.3% are acidophile, 2.2% neutro-baziphile, and the rest of 36.2% are adapted to large variations of soil reaction (Porumb 2006a, b) (Fig. 8.4).

According to the study realized by Bodescu et al. (2010), that was focused on the inventory and mapping of biodiversity in the Cuejdel Lake Protected Area, the floristic spectrum in the studied area contains 161 species, with some taxons already mentioned on the National Red List (Oltean et al. 1994): Liverwort—*Hepatica transsilvanica* (*Ranunculaceae*), endemic but not on the brink of extinction; mountain marigold—*Ranunculus carpaticus* (*Ranunculaceae*), a rare and subendemic species; marsh flower—*Epipactis helleborine* (*Orchidaceaea*), a rare species, etc.

From the rare species found in the area of Stânișoarei Mountains, including on the Cuejdel creek valley and hydrographic basin of Cuejdi, some can be named: chimes flower—*Campanula patula* L. subsp. *abietina* (Campanulaceae); mountain chimes—*Campanula serrata* (Campanulaceae); goiter dove—*Silene nutans* L. subsp. *dubia* (Caryophyllaceae); mountain carnations—*Dianthus collinus* (Caryophyllaceae); forest orchid—*Cephalanthera longifolia* (Orchidaceae); grass snake—*Dactylorhiza majalis* (Orchidaceae); tansy—*Tanacetum macrophyllum* (Asteraceae) ș.a. (Oprea and Sârbu 2008; Bodescu et al. 2010) (Table 8.3).

Aquatic and paludous phytocoenoses. The transition from the forest environment to the water body, especially in the emerged alluvial cone and bank sections specific for wet zones, is dominated by a paludous vegetation represented through a series of Danube communities with: *Eleocharis palustris*, *Typha angustifolia*, *Typha latifolia*, *Juncus effusus*, *Tussilago farfara*, *Schoenoplectus lacustris* or *Scirpus sylvaticus*. At the water surface, the highest spreading is the cluster with *Potamogeton pectinatus* which, during the vegetation season, covers significant surfaces from the tail of the lake and from the shallow sections along the banks and in the barrage area. In total, 11 aquatic and paludous phytocoenosis specific for wet zones were identified surrounding the Cuejdel Lake (Fig. 8.5).

The cluster *Eleocharitetum palustris* (Senikov 19) (Syn.: *Alismato–Eleocharidetum*, Máthé et Kovács 67) is highly distributed along the Cuejdel creek, upstream of the tail of the lake and in the vicinity of the old release area. Horsetail (*Equisetum palustre*), the dominant species of the phytocoenosis is especially developed in the soils with humidity excess and the water level is lower than 10–20 cm. It prefers a marshy type of soil that dries at the surface level significantly during the warm season in the area which marks the transition from the paludous vegetation to the mezophile vegetation, where other species grow with a wider spectrum concerning the exigence towards the hydric factor (Seghedin 1989; Sanda et al. 2008).

The phytocoensosis *Juncetum effusis* (Soó (31) 49) (Syn.: *Epilobio palustri–Juncetum effusi*, Oberdorfer (53) 57; Syn. *Epilobio–Juncetum effusi*, *Mentho–aquatice–Juncetum* effusi, auct. roman.) is composed predominantly of two species: sedge (*Juncus effusus*) and water mint (*Mentha aquatica*). They grow along the Făgetu, Cuejdel and Glodu creeks, as well as sporadically on the lake shores, where there is an excess of humidity and the terrain is marshy and almost flat. The two species form thick clusters and spread along the creeks and upstream of the lake, up to the limit of the middle mountain area (Orberdorfer 1950; Bodescu et al. 2010).

Juncetum inflexis–Menthetum longifoliae (Lohmeyer 53), represented by thick clusters of bulrush (*Juncus inflexus*), appears in islands at the lake tail and along

Table 8.3 Rare plant species from the Cuejdiu valley and Cuejdel Lake area (Oprea and Sârbu 2008; Bodescu et al. 2010; Mihu-Pintilie et al. 2014)

No.	Family	Species	Sozoological category	Int./Nat. regulations	Endemic status in Romania	Romania red lists	Place
1.	*Campanulaceae*	*Campanula patula* L. subsp. *Abietina* (Griseb.) Simonk.	R	BC	Almost endemic	–	Cuejdiu river basin
2.	*Campanulaceae*	*Campanula serrata* (Kit.) Hendrych	R	HD (App.IIb, IVb)	Almost endemic	a	Cuejdiu river basin
3.	*Ranunculaceae*	*Hepatica transsilvanica* Fuss	C	–	Almost endemic	ntR	Cuejdel Lake
4.	*Caryophyllaceae*	*Silene nutans* L. subsp. *Dubia* (Herb.) Zapał.	C	–	Almost endemic	RR	Cozla Hill
5.	*Violaceae*	*Viola jooi* Janka	V	–	Almost endemic	RR	Cozla Hill
6.	*Pinaceae*	*Abies alba* Mill.	E	BC	–	ER	Cuejdiu river basin
7.	*Orchidaceae*	*Cephalanthera longifolia* (L.) Fritsch	nt	–	–	ntR	Cuejdiu river basin
8.	*Asteraceae*	*Cirsium furiens* Griseb. and Schenk	C	BC	Almost endemic	ntR	Piatra Neamţ Village
9.	*Orchidaceae*	*Dactylorhiza majalis* (Rchb.) P. H. Hunt and Summerh.	V	–	–	RR	Cuejdi and Gârcina Villages
10.	*Caryophyllaceae*	*Dianthus collinus* Waldst. and Kit. subsp. *Glabriusculus* (Kit.) Thaisz	V	–	–	RR	Cuejdi and Gârcina Villages
11.	*Caryophyllaceae*	*Dianthus superbus* (L.) subsp. *Alpestris* Čelak.	V	–	–	RR	Muncelu Summit
12.	*Orchidaceae*	*Epipactis helleborine* (L.) Crantz	V	–	–	RR	Cuejdel Lake
13.	*Orchidaceae*	*Epipactis palustris* (L.) Crantz	V	–	–	RR	Cuejdel Lake
14.	*Orchidaceae*	*Epipactis purpurata* Sm.	V	–	–	RR	Piatra Neamţ Village
15.	*Monotropaceae*	*Monotropa hypophegea* Wallr.	V	–	–	RR	Piatra Neamţ Village
16.	*Orchidaceae*	*Neottia nidusavis* (L.) Rich.	V	–	–	RR	Cuejdiu Valley

(continued)

Table 8.3 (continued)

No.	Family	Species	Sozoological category	Int./Nat. regulations	Endemic status in Romania	Romania red lists	Place
17.	*Boraginaceae*	*Omphalodes scorpioides* (Haen.) Schrank	V	–	–	RR	Cozla Hill
18.	*Orchidaceae*	*Orchis coriophora* (L.) subsp. *coriophora*	V	–	–	RR	Cozla Hill
19.	*Orchidaceae*	*Orchis militaris* (L.)	V	–	–	RR	Cozla Hill
20.	*Orchidaceae*	*Platanthera bifolia* (L.) Rich.	V	–	–	RR	Cozla Hill
21.	*Polygonaceae*	*Rumex arifolius* All. (*R. alpestris* Jacq.) f. *carpaticus* Zapał.	nt	–	–	–	Cuejdiu river basin
22.	*Sparganiaceae*	*Sparganium minimum* Wallr.	V	–	–	RR	Cuejdel Lake
23.	*Asteraceae*	*Tanacetum macrophyllum* (Waldst. and Kit.) Sch. Bip.	V	BC	–	RR	Cuejdiu Valley
24.	*Scrophulariaceae*	*Veronica catenata* Pennell	V	–	–	RR	Cuejdiu Valley
25.	*Poaceae*	*Bromus riparius* Rehmann	–	–	–	–	Cuejdiu Valley
26.	*Callitrichaceae*	*Callitriche cophocarpa* Sendtn.	–	–	–	–	Ponoare Hill
27.	*Asteraceae*	*Carpesium cernuum* (L.)	–	–	–	–	Cozla Hill
28.	*Oenotheraceae*	*Epilobium dodonaei* Vill.	–	–	–	–	Piatra Neamţ Village

Sozoological category after Oprea and Sârbu (2008): *R* Red list; *C* Critically endangered plants list; *E* Endemic plants list; *V* Vulnerable plants list; *nt* unthreatening plants. Romanian Red Lists after Oltean et al. (1994): *RR* Rare plant species in Romania; *ER* Endemic plant species in Romania; *ntR* unthreatening plant species in Romania. International regulations: *BC* Bern Convention, 1979; *HD* Habitat Directive (92/43/EEC); Romanian regulations: [a]OUG 57/2007

the Cuejdel creek. Most of the time it is mixed and it is even mistaken with the phytocoenosis *Juncetum tenuis* (Schwik 44) (Syn.: *Juncetum macri*, Diemont et al. 40; Tüxen 50), identified by the presence of American bulrush (*Juncus teunis*). The latter prefers the more trodden lands on the left bank, between the mouths of Făgetu and Glodu creeks (Oltean et al. 1994).

The vegetal association with *Sparganium erectum* (Bur reed) (*Sparganietum erecti*, Roll 38) populates the shallow waters (0.3–0.5 cm) and that are rich in nutrient substances. It prefers the areas from the tail of the lake and from the old release area, where the alluvial accumulation rhythm is high and the water decreases in the warm

Fig. 8.5 Aquatic and paludous phytocoenosis at the mouth of the main tributary of lake

season. In the peak vegetation period it covers approximately 35–85%, percentage that varies depending on the sapropelic qualities of the substrate (Sanda et al. 2008).

The next signaled phytocoenosis is the type *Poo compressae–Tussilaginetum farfarae* (Tüxen 31), with two marker species: coltsfoot (*Tussilago farfara*) and meadow grass (*Poa compressa*). This new association installs itself frequently on landslides, bank erosion and sandy soils, with shade and humidity. It is often encountered on the more steep shores, affected by lacustrine erosion. The floristic component is rich and heterogeneous, with other species that belong to other types of vegetation (meadows, shrubberies, forests).

Almost in the same conditions it develops around the lake the association *Trifolio repenti–Lolietum* (Krippelová 67, Resmeriță and Pop 67) (Syn.: *Lolio–Trifolietum repentis*, Răvăruț et al. 56), identified through the high density of white clover (*Trifolio repentis*). Still, it prefers the mezophile and mezo-higlophile areas in close vicinity of the water body, its vegetation period being conditioned by the light and wet soils (Resmeliță and Pop 1972).

A very important role in supporting the aquatic biodiversity from the coastline of Cuejdel Lake is played by the associations of *Typhetum angustifoliae* și *latifoliae* (Pignatti 53), where the reed with narrow leaf (*Typha angustifolia*) and the reed with wide leaf (*Typha latifolia*) are to be found. Their coenosis is anchored in the vicinity of tail of the lake shores, on the coastline limited by the barrage, as well on the shallow section, no more than 0.5–0.8 cm. Close to them, the silting is more intense due to the rhizomes network collecting the alluvial material and organic waste. This phenomenon is underlined at the mouths of Glodu and Cuejdel creeks and in the area of the alluvial bank from the old release system. Within this association can appear other higrophile species, such as *Bidention* and *Agropyro–Rumicion* (Sanda and Popescu 1997; Sanda et al. 2008).

Still from the aquatic phytocoenosis is the association *Schoenoplectetum lacustris* (Chouchard 24) with round reed (*Schoenoplectetus lacustris*). The dominant species reaches the development optimum in the areas with water depth of 70–80 cm in the outlet area of the Făgetu, Cuejdel, Glodu and Pârâul Rotund creeks, at the tail of the lake and on the left bank of the barrage. It can also develop for a longer time on land and is also to be found upstream from the lake or on the main tributaries flows. Similarly the vegetal association *Scirpetum sylvatici* (Ralski 31, Maloch 35 em, Schwick 44), a phytocoenosis cleared by the marker species *Scirpus sylvaticus* (forest bulrush) will also develop. This plant is specific for the lower mountain zone from the Eastern Carpathians and can be found in an abundant density on the shores of Cuejdel Lake. It is dependent on alluvial soils with humidity excess, which favors development of other higlophile and mezophile species through vegetal alliance (*Festuca rubra*) (Table 8.4).

The species that provide most organic substance are *Potamogeton pectinatus* (91.29%/100 g) and *Thypa latifolia* (90.04%/100 g). Also they have the most spread out phytocoenosis on the aquatic surface, contributing to the silting of limitrophe areas from the lake tail through direct decomposition or retention of alluvial material. The consistent mineral share is given by the presence of species *Chara fragilis* (44.74%/100 g) and *Carex remota* (42.17%/100 g), while the highest quantity of rough proteine is given by *Lemna minor* (18.30%/100 g) and *Thypa latifolia* (19.20%/100 g). Nevertheless, the reduced frequency of surfaces with hydrophile vegetation surrounding the lake makes the biomass production remain small compared to the vegetation potential. In addition to organic substances the trees from the water body are also important; the water formed the lake and supplies it or retains the organic load on the lacustrine depression (Table 8.5).

The macrophyte species that are representative within the aquatic and palustrine phytocoenoses around the Cuejdel Lake hold various contents of nutrients (proteins, sugars, minerals etc.). The vegetal biomass analysis that was performed on around 10 species of algae and superior plants clearly illustrates their eco-biochemical identity. The results could be correlated with the spatial distribution and density of the analysed species, thus providing a clear picture of the biomass production areas (Grabowski 1973; Mititelu et al. 1992) (Fig. 8.5).

The ultra-hydrophyte phytocenosis with the highest density in the lake area is represented by *Potamogetonetum pectinati* (Carstensen 55). This is mainly prevalent in the shore area (−0.2 to 1.5 m depth) and in the upper lake area, where the lacustrine bottom is mostly covered in typical limnic soil with peat insertions. Fennel pondweed (*Potamogeton pectinatus),* the prevalent species, takes up to 80% and is usually accompanied by several other natant species (*Potamogeton crispus, P. nodosus, Myriophyllum spicatum*). Together, they make up for a large part of the biomass, thus contributing to the silting of the lacustrine where they grow. As far as the Cuejdel Lake is concerned, this phytocenosis is not largely spread, but influences

Table 8.4 Wetlands plant associations corresponding to Cuejdel Lake (Mihu-Pintilie et al. 2014)

Association	Dominant species	Other species	Place
Eleocharitetum palustris Senikov 19 Syn.: *Alismato-Eleocharidetum* Máthé et Kovács 67	*Equisetum palustre*	*Oenante aquatica, Mentha aquatica, Phragmites australis, Berula erecta, Agrostis stolonifera, Potentilla reptans.*	Cuejdel Valley, upper lacustrine basin and natural dam area
Juncetum effusis Soó (31) 49 Syn.: *Epilobio palustri-Juncetum effusi* Oberdorfer (53) 57; Syn. *Epilobio-Juncetum effusi, Mentho-aquatice-Juncetum effusi* auct. roman.	*Juncus effusus* *Mentha aquatica*	*Myosotis scorpioides, Scirpus sylvaticus, Lysimachia vulgaris, Deschampsia cespitosa, Mentha longifolia, Lycopus europaeus, Lychnis flos-cuculi, Carex dinstans.*	River mouths: Făgetu, Cuejdel and Glodu brooks
Juncetum inflexis-Menthetum longifoliae Lohmeyer 53	*Juncus inflexus*	*Agrostis stolonifera, Ranunculus repens, Potentilla reptans, Carex hirta, Plantago major, Trifolium repens, Prunella vulgaris.*	Cuejdel Valley and upper lacustrine basin
Juncetum tenuis Schwik 44 Syn.: *Juncetum macri* Diemont et al. 40, R. Tüxen 50	*Juncus tenuis (macer)*	*Lolium perenne, Poa annua, Plantagomajor, Carex hirta, Ranunculus repens, Trifolium repens, Prunella vulgaris, Taraxacum officinale, Trigonella procumbens*	Left bank of Cuejdel Lake. River mouths: Făgetu and Cuejdel brooks
Sparganietum erecti Roll 38	*Sparganium rectum*	*Phragmition communis—Berula erecta, Butomus umbellatus, Calystegia sepium, Lycopus europaeus, Phragmites australis, Rumex hydrolapathum, Sagittaria sagittifolia, Schoenoplectus lacustris, Sium latifolium, Sparganium erectum, Typha schuttleworthii.*	Cuejdel Valley, upper lacustrine basin and natural dam area
Poo compressae-Tussilaginetum farfarae R. Tüxen 31	*Tussilago farfara* *Poa compressa*	*Artemisietea vulgaris—Daucus carota, Melilotus officinalis, Rumex obtusifolius, Cirsium vulgare, C. Arvense, Equisetum arvenses, Mentha longifolia, Taraxacum officinale, Carduus personatus.*	Upper and left bank of Cuejdel Lake and river mouth of Făgetu, Cuejdel and Glodu brooks
Trifolio repenti-Lolietum Krippelová 67, Resmeriță and Pop 67 Syn.: *Lolio-Trifolietum repentis* Răvăruț et al. 56	*Trifolio repentis*	*Poa pratensis, Prunella vulgaris, Plantago lanceolata, Achillea millefolium, Medicago lupulina, Taraxacum officinale, Lotus corniculatus.*	Large area of Cuejdel Lake
Typhetum angustifoliae/latifoliae Pignatti 53	*Typha angustifolia/latifolia*	*Bidention- Agropyro-Rumicion—Scirpo-Phragmitetum—Phragmites australis, Alisma plantago-aquatica, Oenanthe aquatica, Butomus umbellatus.*	Upper and left bank of Cuejdel Lake and river mouth of Pârâul Rotund, Piciorul Crucii, Făgetu, Cuejdel and Glodu brooks

(continued)

Table 8.4 (continued)

Association	Dominant species	Other species	Place
Schoenoplectetum lacustris Chouchard 24	*Schoenoplectetus lacustris*	*Typha augustifolia, Glyceria maxima, Phragmites australis, Rumex hydrolapatum, Butomus umbellatus, Cladium mariscus.*	Upper and left bank of Cuejdel Lake and river mouth of Făgetu, Cuejdel and Glodu brooks
Scirpetum sylvatici Ralski 31, Maloch 35 em. Schwick 44	*Scirpus sylvaticus*	*Caltha palustris* ssp. *Laeta, Carex acuta, Chaerophyllum hirsutum, Equisetum palustre, Juncus effuses, Poa palustris, Prunella vulgaris, Trifolium repens.*	Cuejdel and Glodu Valleys, upper lacustrine basin and natural dam area
Potamogetonetum pectinati Carstensen 55	*Potamogeton pectinatus*	*Potamogeton crispus, P. nodosus, Myriophyllum spicatum, Ranunculus tricophyllus, R. aquatilis, Ceratophyllum demersum.*	Upper lacustrine basin and littoral zone with depths >1.5 [m]

Table 8.5 Biochemical characteristics of aquatic and paludous macrophytes from the lacustrine wetlands of Cuejdel Lake (Grabowski 1973; Mititelu et al. 1992; Bodescu et al. 2010)

Species	[%] from 100 g of dry substance		
	Organic substance	Mineral substance	Total protein
1. *Alisma plantago aquatica*	85.74	14.16	24.62
2. *Chara fragilis*	55.26	44.74	2.27
3. *Carex remota*	57.83	42.17	3.90
4. *Equisetum fluviatilis*	85.78	14.22	13.81
5. *Lemna minor*	77.50	22.50	18.30
6. *Mentha aquatica*	77.47	22.53	5.61
7. *Myriophyllum spicatum*	86.79	13.21	10.80
8. *Potamogeton pectinatus*	91.29	8.71	10.48
9. *Ranunculus repens*	88.43	11.57	18.03
10. *Typha latifolia*	90.04	9.96	19.20

the water oxygenation regime during summer and provides the organic material for the productive bottom area of the lake tail (Mititelu et al. 1992).

Phytoplankton. The purpose of plankton reasearch was the qualitative depiction of the trophic chains within the aquatic environment, based on primary producers. The algae component was determined through sampling from various areas of the lake (tributary rivers, input, the shore area etc.), to which the main tributaries' outlets were added. The water samples were collected with a Nansen bottle and filtered

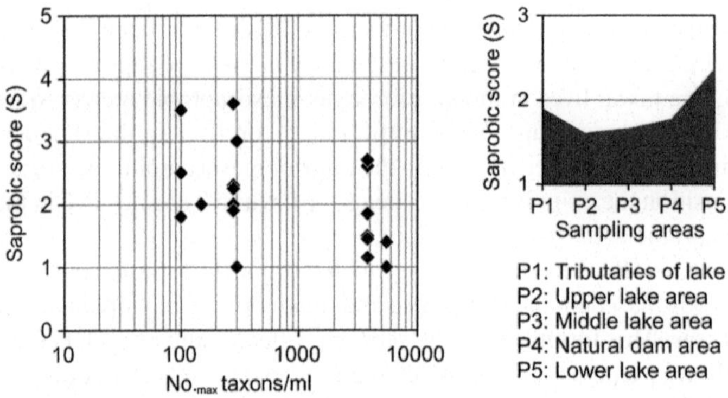

Fig. 8.6 The saprobic score (S) based on the abundance and quality of the phytoplankton in the five sampling areas from Cuejdel Lake

through a closing net, in situ. The results, relevant as for 2013, were correlated with similar studies conducted between 2004 and 2008 (Porumb 2006a, b).

The results reveal a relatively well defined phytoplankton biodiversity. An overall 50 taxons were identified, out of which 41 species were identified between 2004 and 2008 (Porumb 2006a, b), with nine new species identified in 2013, throughout our own research. The phytoplankton species belong to six systemic algae groups: *Cyanophyta, Chrysophyta, Bacillariophyta, Pyrrophyt, Chlorophyta* și *Euglenophyta.*

The dominant systemic category was *Bacillariophytai*, with *Cyclotella ocellata, C. comta* and *Achnanthes minutissima* as dominant species, followed by *Chrysophyta* and *Euglenophyta.* The *Cyanophyta* group is significantly present in the outlet areas of the tributary rivers, as well as in the dam area, which is proof of the relative contamination of the upstream waters due to antrophic activity. The *Chlorophyta* group algae holds a significant share, with their highest prevalence in the center area of the lake, due to longer periods of insolation.

According to the location of sampling areas and the month when the analyses were carried out, a significant difference in the total number of algae was noticed. The water thermal regime determines the following variations of the average algae number: inverted stratification (in winter) 500–550 items/ml, spring homeothermy 1000–1100 items/ml, direct stratification (summer) 1250–1300 items/ml, autumn thermal stratification 900–950 items/ml. As far as the systemic groups are concerned, the phytoplankton in the samples collected ranges between the following variations: *Cyanophyta* 50–100 items/ml, *Chrysophyta* 70–5500 items/ml, *Bacillariophyta* 780–3800 items/ml, *Pyrrophyt* 60–300 items/ml, *Chlorophyta* 55–280 items/ml and *Euglenophyta* 45–150 items/ml (Fig. 8.6).

The algae group distribution has a high monthly fluctuation, with a high prevalence of diatoms in July (*Bacillariophyta)* and of the *Pyrrophyta* algae in August. This behaviour is proof that phytoplankton is highly sensitive and indicative of the changes in the environment.

Some of the algae found have the property to describe the saprobic nature of the aquatic environment. Out of the 24 representative taxons, the saprobic distribution is as follows: one taxon lives in the inferior xeno–mesosaprobic area (x–α), 3 species live in the superior mesosaprobic area (β), one taxon lives in the superior–inferior transition mesosaprobic area (β–α), 4 in the inferior mesosaprobic area (α), one in the inferior transition mezo–polysaprobic area (α–p), while 2 species live in the inferior polymesosaprobic area (p–α) (Table 8.6).

The *Pantle and Buck* method was used to determine the saprobic index in the Cuejdel water body, based on the abundence and quality of the phytoplankton in the five sampling areas. This formula uses the saprobic valence sum of each taxon (S), multiplied with the frequency of the indicating species (1—scarce frequency; 2—high frequency; 3—mass development). Thus, the outlet of the main lake tributaries (P1) holds a 1.9 saprobic score, which is the equivalent of an area of medium impurification (the superior β–mesosaprobic area). The superior (P2) and central (P3) areas of the lake hold a saprobic index ranging between 1.61 and 1.66, which is a sign of low to medium impurification (the superior oβ–oligo–mesosaprobic area). The dam area (P4) holds a 1.77 score, corresponding to the value in the (P1) area, while in the (P5) sampling area, a value of 2.36 indicates a medium to high impurification due to the debris in the outlet passage (the inferior–superior βα–mesosaprobic area) (Table 8.6).

In conclusion, the structure and distribution of the phytoplankton in the Cuejdel Lake indicates that algae biodiversity is in its incipient stage of development. Also, according to M. No. 161/2006 on determining the trophic level based on the lacustrine ecological bioindicators, the correspondence between the saprobic valence location and the trophicity of the water is representative for the oligotrophic to mezotrophic nature of the ecosystem (Porumb 2006a, 2006b; Oprea and Sârbu 2008; Bodescu et al. 2010; Mihu-Pintilie et al. 2014)

8.2 Fauna and Ecological Bio Indices in the Water Body

Given the vertebrate and invertebrate biodiversity, fauna plays an important part in the Cuejdel Lake ecosystem structure. Among the various vertebrate species, we mention birds, reptiles, batrachians and fish, while for the invertebrate species, a series of worms, mollusks and arthrophods (arachnids, crustaceans, myriapods and inspects) are the most representative. The zooplankton and macroinvertebrate water body benthic and near-shore benthic communities are also to be mentioned.

The vertebrate fauna. The mammals living in the lake area are specific to mixed forests, some of them preferring the vicinity of the lacustrine basins because the food offer is richer. These include the hedgehog (*Erinaceus europaeus* L.), the alpine shrew (*Sorex alpinus* Sch.), the European hare (*Lepus europaeus* Pall.), the red squirrel (*Sciurus vulgaris* L.), the field vole (*Microtus agrestis* L.), the wild boar (*Sus scropha* L.), the brown bear (*Ursus arctos*), as well as the grey wolf (*Canis lupus*).

Table 8.6 The saprobic score for each algae species determined in Cuejdel lacustrine basin and main tributaries of lake (Mihu-Pintilie et al. 2014)

Systemic group/Taxon	P1	P2	P3	P4	P5	(s)	(g)	(S)
CYANOPHYTA (50–100 ex./ml)								
Microcystis pulverea	•	•	·	•+	·	β	3	1.8
Oscillatoria sp.	+	·	·	·	·	α	4	2.5
Spirulina laxissima	+	·	·	·	·	α–p	3	3.5
CHRYSOPHYTA (70–5500 ex./ml)								
Chromulina sp.	+	+	+	·	·	o	5	1.0
Chrysococcus rufescens	·	+	+	·	·	o–β	3	1.4
Ochromonas sp.	·	·	+	·	·	–	–	–
Kephyryon ovum	·	·	·	·	+	–	–	–
BACILLARIOPHYTA (780–3800 ex./ml)								
Cyclotella comta	•	•	•	•	+•	o–β	3	1.15
Cyclotella ocellata	•	+•	+•	+•	+•	–	–	–
Cyclotella meneghiniana	•	·	·	·	•	β–α	3	2.6
Diatoma elongatum	+	·	·	·	·	o–β	3	1.5
Achnanthes affinis	+	+	·	·	·	–	–	–
Achnanthes minutissima	+	+	+	+	·	o–β	2	1.45
Achnanthes sp.	+	·	·	·	·	–	–	–
Navicula cryptocephala	·	·	·	·	+	α	4	2.7
Navicula sp.	·	·	+	·	·	–	–	–
Cymbella microcephala	·	+	·	·	·	–	–	–
Cymbella tumida	+	·	·	·	·	–	–	–
Cymbella affinis	·	•				–	–	–
Cymbella sp.	·	·	+	·	·	–	–	–
Gomphonema olivaceum	+		•	·	·	x–α	1	1.85
Nitzschia palea	+	+•	·	·	·	α	3	2.7
Nitzschia sp.	+	+	·	·	·	–	–	–
Surirella ovata	·	+	·	·	·	β	2	1.85
Melosira sp.	·	·	·	•	·	o–β	4	1.85
PYRROPHYTA (60–300 ex./ml)								
Chroomonas acuta	·	+	+	+	·	–	–	–
Chroomonas nordstedtii	·	+	+		·	–	–	–
Chroomonas minima	·		+	+	·	–	–	–
Chroomonas sp.	·	+	·	·	·	–	–	–
Rhodomonas pusilla	·	·	·	·	+	–	–	–
Cryptomonas ovata	·	·	·	·	+	α	4	3.0
Cryptomonas marssonii	·	·	·	·	+	–	–	–

(continued)

Table 8.6 (continued)

Systemic group/Taxon	P1	P2	P3	P4	P5	(s)	(g)	(S)
Cryptomonas sp.	·	·	·	·	+	–	–	–
Gymnodinium sp.	·	+	·	+•	·	o–β	–	–
Peridinium cinctum	·	·	+	·	·	o	5	1.0
Perdinium sp.	·	·	•	·	·	o	5	1.0
Glenodinium cinctum	·	•	•	·	·	–	–	–
Glenodinium sp.	·	·	·	+	·	–	–	–
CHLOROPHYTA (55–280 ex./ml)								
Chlamydomonas sp.	·	·	•	•	+•	β	4	2.3
Pandorina morum	·	+				β	3	2.0
Chodatella sp.	·	·	+	+	+	–	–	–
Chlorella vulgaris	·	·	+•	+	+	p–α	3	3.6
Ankistrodesmus bibraianum	+	·	·	·	·	β	3	2.25
Keratococcus sp.		+	·	·	·	–	–	–
Scenedesmus acutus	+	+	·	·	·	β	5	1.9
Scenedesmus sp.	•		·	·	·	–	–	–
EUGLENOPHYTA (45–150 ex./ml)								
Trachelomonas volvocina	•	+	+•	·	+	β	3	2.0
Trachelomonas acanthostoma	•	•	•	•	•	–	–	–
Trachelomonas planctonica	·	·	·	•	·	–	–	–
Tribonema minus	·	·	·	•	·	–	–	–

P1, P2, P3, P4 and P5: Sampling areas [P1—Tributaries of lake; P2—Upper lake area; P3—Middle lake area; P4—Natural dam area; P5—Lower lake area and outlet]; + Species determined between 2004–2008; • Species determined between 2012–2013. (s): Saprobic classes [o—oligosaprobic area; β—superior mesosaprobic area; β–α—superior-inferior transition mesosaprobic area; α—inferior mesosaprobic area; α–p—inferior transition mezo-polysaprobic area; p—polymesosaprobic area; p–α—inferior polymesosaprobic area; x–α—xeno-mesosaprobic area]; (g): Saprobic weight; (S): Saprobic score

The bird fauna is also characteristic for mixed forests, but the aquatic and palustrine species are few. Among the 37 identified species, the wild duck (*Anas platyrhynchos*) has frequently been noticed in the upstream wetland area, pair nesting. It is most likely that it is a mere winter guest in the area or it might be migrating between the lakes in the Bistriţa Valley (Izvorul Muntelui, Bâtca Doamnei, Vaduri etc.) and the small lakes in the mountain region (Mihu-Pintilie et al. 2014).

Overall, 18 species are sedentary (ex. *Certhia familiaris*), out of which only 3 stay here for the winter (ex. *Accipiter nisus*), 5 species are partially migratory (ex. *Buteo buteo*), 14 species spend summers only in this region (ex. *Phoenicurus ochoruros*), out of which 3 rarely spend the winter here, as well, and one species only transits the area (*Phylloscopus trochilus*). The birds included in the 1st Appendix of the Habitats Directive are protected, with 2 species of community interest (*Bonasa*

Fig. 8.7 Carpathian newt (*Lissotriton montandoni* L.; Syn.: *Triturus montandoni*—Boulenger 1860), an endemic species in Eastern Carpathian identified on the littoral zone of **a** Cuejdel Lake and **b** Blue lake; **c** European common frog (*Rana temporaria* L., Syn.: *R. Aragonensis*—Planca Soler, Rodriguez Vieites and Suarez Martinez 1995) in the wetlands of Cuejdel and Glodu creeks; **d** Sand lizard (*Lacerta agilis* L., Syn.: L. *boemica*—Suchov 1929) sitting on a fossil trunk in Fagetu creek area; **e** Yellow-bellied toad (*Bombina variegate* L., Syn.: *Rana variegata*—Linnaeus 1758) in the wetlands of Piciorul Crucii creek

bonasia, Dendrocopos syriacus), 10 species of national interest which are included in Appendix 4F of Directive no. 57/2007 (ex. *Erithacus rubecula*), while 2 other species are part of the Red National List (*Corvus corax, Upupa epops*) (Munteanu 2009) (Table 8.7).

The Cuejdel Lake and the corresponding wetland areas are an ideal aquatic ecosystem for amphibians and reptiles. At the same time, the mixed forests nearby ensure suitable habitats for these species. A total of 11 taxons have been identified, with 7 of them being widely spread in the area: *Salamandra salamandra, Lissotriton montandoni, Ichthyosaura alpestris, Bombina variegate, Rana temporaria, Lacerta agilis, Zootoca vivipara*, while 4 of them are more narrowly distributed: *Triturus cristatus, Lissotriton vulgaris, Pelophylax ridibundus* and *Natrix natrix*. The presence of 3 other species has been signalled in the area, but these have not yet been identified: *Hyla arborea, Anguis fragilis* and *Vipera berus* (Fig. 8.7).

The lacustrine ichthyofauna is represented by the Pisces sub-branch (*Osteichthyes* Class), of which we mention the brown trout (*Salmo trutta fario* L.). In the first years after the lake was formed, the density of this species was rather high, whereas rare captures have been made lately. Man-introduced species include high densities of crucian carps (*Carassius carassius* L.), the European perch (*Perca fluviatilis* L.)

Table 8.7 Bird species list from the Cuejdel Lake area (Mihu-Pintilie et al. 2014)

Family	Species	Common name	Phenolog. cat.	Int./Nat. regulations		
				BD	OUG 57/2007	RRL
Accipitridae	*Accipiter nisus*	Sparrowhawk	S/W	–	–	–
Anatidae	*Anas platyrhynchos*	Wild duck	Pm/W	App.III/1	App.5C,5D	–
Tetraonide	*Bonasa bonasia*	Hazel grouse	S	App.II/2	App.5C	–
Accipitridae	*Buteo buteo*	Buzzard	Pm	–	–	–
Fringillide	*Carduelis carduelis*	Goldfinch	S/W	–	App.4B	–
Certhiide	*Certhia familiaris*	Treecreeper	S	–	–	–
Fringillide	*Coccothraustes coccothraustes*	Hawfinch	S	–	–	–
Columbide	*Columba palumbus*	Wood pigeon	W/Wr	App.II/1, III/1	App.5C,5D	–
Corvide	*Corvus corax*	Raven	S	–	App.4B	E
Cuculide	*Cuculus canorus*	Cuckoo	Sb	–	–	–
Hirundinide	*Delichon urbica*	House martin	Sb	–	–	–
Picide	*Dendrocopos major*	Great spotted woodpecker	S	–	–	–
Picide	*Dendrocopos syriacus*	Syrian woodpecker	S	App.I	App.3	–
Turdide	*Erithacus rubecula*	European robin	Sb/Wr	–	App.4B	–
Fringillide	*Fringilla coelebs*	Chaffinch	Pm	–	–	–
Corvide	*Garrulus glandarius*	Eurasian jay	S	App.II/2	App.5C	–
Fringillide	*Loxia curvirostra*	Red crossbill	S	–	–	–
Motacillide	*Motacilla alba*	White wagtail	Sb	–	App.4B	–
Motacillide	*Motacilla cinerea*	Grey wagtail	Sb/Wr	–	App.4B	–
Corvide	*Nucifraga caryocatactes*	Nutcracker	S	–	–	–
Paride	*Parus ater*	Coal tit	S	–	–	–
Paride	*Parus caeruleus*	Blue tit	S	–	–	–

(continued)

Table 8.7 (continued)

Family	Species	Common name	Phenolog. cat.	Int./Nat. regulations		
				BD	OUG 57/2007	RRL
Paride	*Parus major*	Great tit	S	–	–	–
Paride	*Parus montanus*	Willow tit	S	–	–	–
Paride	*Parus palustris*	Marsh tit	S	–	–	–
Turdide	*Phoenicurus ochruros*	Black redstart	Sb	–	App.4B	–
Turdide	*Phylloscopus collybita*	Chiffchaff	Sb	–	App.4B	–
Turdide	*Phylloscopus trochilus*	Willow warbler	Pb/Sb	–	App.4B	–
Fringillide	*Pyrrhula pyrrhula*	Bullfinch	S	–	–	–
Turdide	*Regulus regulus*	Goldcrest	Pm/W	–	App.4B	–
Turdide	*Saxicola rubetra*	Whinchat	Sb	–	–	–
Sittide	*Sitta europaea*	Wood nuthatch	S	–	–	–
Sylvide	*Sylvia atricapilla*	Blackcap	Sb	–	–	–
Troglodytide	*Troglodytes troglodytes*	Wren	Sb/Wr	–	–	–
Turdide	*Turdus merula*	Blackbird	Pm	–	–	–
Turdide	*Turdus philomelos*	Song thrush	Sb	–	App.5C	–
Upupide	*Upupa epops*	Hoopoe	Sb	–	App.4B	V

Phenolog. cat. (Phenological categories): *S* Sedentary bird species; *Pm* Partially migratory bird species; *W* Wintering bird species; *Sb* Summer bird species; *Wr* Wintering rare bird species; *Pb* Passage bird species; *BD* Birds Directive (79/409/EEC); RRL: Romania Red Lists of bird species: *E* Endangered bird species; *V* Vulnerable bird species

and the pumpkinseed (*Lepomis gibbosus* L.), while the invasive species include the common minnow (*Phoxinus phoxinus* L.) and the gudgeon (*Gobio gobio* L.) (Fig. 8.8 and 8.9).

The invertebrate fauna. The abundance and variety of the invertebrate fauna in the research area is closely related to the microclimate induced by the mixed forests and the water meadows. The most numerous invertebrate group is represented by insects, of which more than 100 have been identified. These are part of the following orders: Odonata, Ortoptera, Heteroptera, Hymenoptera, Coleoptera, Lepidoptera and Diptera.

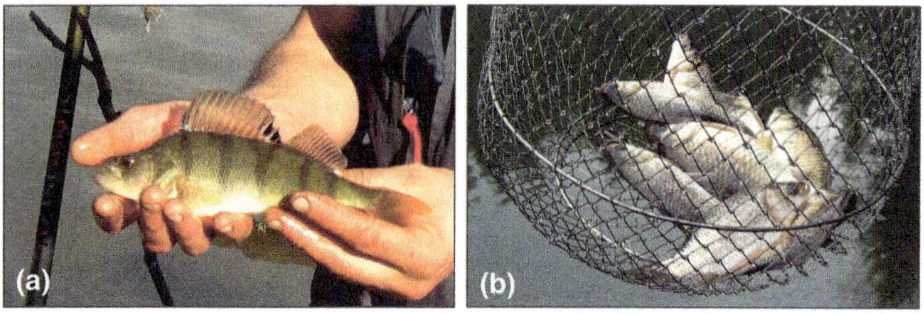

Fig. 8.8 Fish species introduced by man in Cuejdel Lake: **a** European perch (*Perca fluviatilis* L.) and **b** Crucian carps (*Carassius carassius* L.)

Fig. 8.9 Common bird species from Cuejdel Lake area and Stânișoarei Mts.: **a** Hazel grouse (*Tetrastes bonasia*), **b** Western capercaillie (*Tetrao urogallus*), **c** White wagtail (*Motacilla alba*), **d** Ural Owl (*Strix uralensis*), **e** Wild duck (*Anas platyrhynchos*), **f** Great spotted woodpecker (*Dendrocopos major*) (Photograph by C. Ion—Romanian Ornithological Society)

The Odonata order includes a number of carnivorous dragonflies that live around the lake. Some of them catch their prey in flight (*Libellulidae Family)*, while the others feed on aphids or other small insects (*Aeschnidae Family*, *Agrionidae Family, Coenagrioidae Family*). The most representative families of the other orders include: the Ortoptera Order—*Tettigoniidae Family, Decticidae Family, Acrididae Family*; the Order Heteroptera—*Scutelleridae Family, Pentatomidae Family, Coreidae Family, Gerridae Family;* the Hymenoptera Order—*Siricidae Family, Tenthredinidae Family, Braconidae Family, Ichneumonidae Family, Cynipidae Family, Vespidae Family, Apidae Family, Megachilidae Family*; the Coleoptera Order—*Carabidae Family, Cantharidae Family, Elateridae Family, Coccinelidae Family, Scarabei-*

dae Family, Cerambycidae Family, Chrysomelidae Family, Ipidae Family; the Lepidoptera Order—*Pieridae Family, Nymphalidae Family, Satyridae Family*; the Diptera Order—*Tachinidae Family, Syrphidae Family, Calliporidae Family, Tabanidae Family, Asilidae Family, Bombyllidae Family*. Part of the larvae of these insects live in the water body and they provide the abundant zooplankton and zoobenthos communities, many of them being food for the fish and birds that live in the lake area.

Several studies have focused on the qualitative and quantitative distribution of the zooplankton communities in the Cuejdel Lake area. Research dating back to 2004–2007 indicates that the zooplankton is mainly made of three systemic groups (Cladocera, Copepoda and Rotatoria), with the following species: *Acanthocyclops viridis, Asplachna priodonta, Bosmina longirostris, Chydorus sphaericus, Eucyclops serrulatus, Filinia maior, Keratella cochlearis, Lepadella ovalis, Pseudochydonus globosus, Polyarthra dolichoptera, P. major, P. remota, P. remota* and *P. vulgaris*.

The highest numeric density is held by the Rotatoria group, with over 70,000 items/m^3 in the summer and over 100,000 items/m^3 in autumn. The Cladocera and Copepoda groups make up 40% of the overall zooplankton share, with a cumulated density of up to 10,000 items/m^3. In 2008, out of the total 172 items/L (0.204 mg/L) of the zooplankton species that were dominant in the warm season, in July, 154 items/L (0.072 mg/L) were part of the dominant Rotatoria group, while in August, 24 items/L (0.008 mg/L) of the total 44 items/L (0.136 mg/L) were part of the same systemic group.

Along with completing the zooplankton species list, our research between 2012 and 2013 aimed to provide the qualitative description of the benthic macroinvertebrate communities (zoobenthos). The benthic samples were collected during the warm season (July–August), as well as during the cold season (February) with a benthic grab sampler (Van Veer grabs—0.5 L). Analyses were carried out on taxons in the laboratory and two major dominant classes were identified: Oligochaeta (*Limnodrilus hoffmeisteri, Potamothrix hammoniensis, Tubifex tubifex*) and Diptera (*Bezzia* sp., *Chaoborus flavicans, Chironomus plumosus, Tanypus punctipennis*) (Speczíár 2008; An et al. 2012; Tarmo 2013).

During the warm season, red worms (Oligochaeta) are the most abundant species with 10–80 items/L (*Tubifex tubifex*—10–50 items/L; *Potamothrix hammoniensis*—2–58 items/L; *Limnodrilus hoffmeisteri*—0–6 items/L), while in the cold season, below the ice sheet, the density drops to 8–28 items/L. The chironomidae larvae are at their most abundant in winter, with 20–300 items/L (*Chaoborus flavicans*—2–298 items/L), while in the summer their density drops to 6–36 items/L.

In total, the number of items varies all year round between 10–310 items/L. According to the depth where the samples were collected, the highest abundance is to be found in the alluvial fan in the lake tail area (0–8 m depth—summer 10–50 itmes/L; winter—50–300 items/L), and in the natural dam area (0–3 m depth—80 items/L in the summer).

The ecological bio indicators. The dominant bio indicators on the vertebrate and invertebrate list that is characteristic for Cuejdel Lake are part of the following orders: Pisces, Insecta (Suborders *Odonata, Heteroptera, Diptera*), Crustacea (Suborders Cladocera, Copepoda), Annelida (Suborders *Oligochaeta*) and Vermes (Suborders

Rotifera). Out of the 23 representative taxons, the distribution on saprobic zones is as follows: 12 belong to the oligosaprobic zone (o), out of which 2 taxons are present in the xeno–oligosaprobic zone (x–o) and 4 species in the upper oligo–mesosaprobic zone (o–β), 7 species live in the upper mesosaprobic zone (β), of which 4 are present in the upper–lower transition mesosaprobic zone (β–α); furthermore, three of the taxons indicate polysaprobic tollerance (p) (Table 8.8).

According to their specific valence (S), 5 taxons hold values between 0.4–1.0 (ultra–oligotrophic), 11 species fall in the 1.1–2.0 interval (oligotrophic), while another 7 taxons have a valence between 2.1 and 4.0 (mesotrophic–eutrophic). According to the bathymetric level where the representative bio indicators are to be found, the saprobic valence is as follows: 0.5–1.15 on the water surface (oligosaprobic), hypolimnion 0.4–2.35 (upper oligomesosaprobic), epilimnion and the benthic zone 1.15–3.8 (oligo–polysaprobic).

According to the ichthyofauna as bio indicator and to the qualitative structure of the zooplankton, the Cuejdel aquatic ecosystem is enclosed in the oligrotrophic and oligo-mesotrophic lake category. Also, the low numeric density of the zooplankton and the identified taxons is proof that the trophic basis of the lake is relatively scarce, therefore providing the oligotrophic nature of the water quality.

As far as the aquatic microorganisms are concerned, previous quantitative research established 12 ecophysiological bacteria groups in the Cuejdel Lake. Their role is important for the circuit of biogenous elements, such as nitrogen, carbon and sulphur, while at the same time pointing towards the water quality. Their development also correlates with the physical and chemical features of the water, as the ammonification, denitrification and aerobic–heterotrophic bacteria indicate the trophic level in the aquatic ecosystem.

For 2008, the quantitative distribution according to bacteria groups and based on the geometric mean was as follows: ammonification bacteria 629–4470 cells/ml, denitrification bacteria 178–891 cells/ml, anaerobic nitrogen–fixing bacteria 5–93 cells/ml, amylolytic bacteria 479–6920 cells/ml, aerobic cellulose bacteria 2–24 cells/ml, anaerobic cellulose bacteria <1–8 cells/ml, aerobic heterotrophic bacteria 871–10,700 cells/ml, decomposing bacteria of sulphurous organic matter 6–32 cells/ml, sulfur reducing bacteria <1–2.6 cells/ml, saprophytes (37 °C) 115–5310 celule/ml, total coliform bacteria 39–324 cells/ml. These values show that aerobic bacteria is poorly represented, meaning that organic matter decomposes and undergoes mineralization mainly in anaerobic conditions (Table 8.8).

The high density of the ammonification, denitrification, amylolytic, aerobic heterotrophic and psychrophile saprophyte bacteria in 2008 placed the Cuejdel aquatorium in the oligotrophic water ecosystem group and in the 1st category of water quality. These results confirm the dominant ecological bio indicators based on trophic level analysis performed between 2012 and 2013. However, a variation towards an oligo–mesotrophic nature was noticed which is mainly prominent during the warm season, when water contamination is at its highest due to human activity.

Table 8.8 The saprobic score of the main ecological bioindicators species of water quality within the Cuejdel lacustrine basin and lake tributaries (Mihu-Pintilie et al. 2014)

Systemic group/species	(s)	x	o	β	α	p	(g)	(S)
PISCES								
Carassius carassius	β–α	·	·	5	5	+	3	2.5
Gobio gobio	β–α	·	2	4	4	+	2	2.25
Perca fluvialitis	β	·	3	6	1		3	1.75
Phoxinus phoxinus	o	2	6	2	·	·	3	1.0
Salmo trutta fario	x–o	6	4	+	·	·	3	0.4
INSECTA								
Odonata—*Aeschna* sp.	x–o	5	5	·	·	·	3	0.5
Heteroptera—*Gerris* sp.	o–β	1	4	4	1	·	1	1.55
Diptera -								
Chaoborus flavicans	o–p	·	2	4	2	2	1	2.25
Chironomus plumosus	p	·	·	+	2	8	4	3.8
Tabanaide sp.	β–α	3	3	2	2	·	1	1.15
CRUSTACEA Cladocera-								
Bosmina longirostris	o–β	1	4	4	1	·	1	1.55
Chydorus sphaericus	β	1	3	4	2	·	1	1.65
Pseudochydonus globosus	o	·	8	2	·	·	4	1.2
Copepoda -								
Acanthocyclops viridis	o	2	5	3	·	·	2	1.15
Eucyclops serrulatus	o–β	·	4	4	2	·	2	1.85
ANNELIDA Oligochaeta-								
Limnodrilus hoffmeisteri	p	·	·	+	3	7	4	3.7
Tubifex tubifex	p	·	·	+	2	8	4	3.8
VERMES Rotifera-								
Asplachna priodonta	o–β	1	4	4	1	·	1	1.55
Filinia maior	β–α	·	1	5	4	·	2	2.35
Keratella cochlearis	β	2	3	5	·	·	2	1.55
Lepadella ovalis	o	1	6	3	·	·	3	1.25
Polyarthra remota	o	·	10	·	·	·	5	1.0
Polyarthra vulgaris	o	·	10	·	·	·	5	1.0

(s): Saprobic classes [o—oligosaprobic area; β—superior mesosaprobic area; β–α—superior-inferior transition mesosaprobic area; α—inferior mesosaprobic area; α–p—inferior transition mezo-polysaprobic area; p—polymesosaprobic area; p–α—inferior polymesosaprobic area; x–α—xeno-mesosaprobic area] and descriptive score for each saprobic classes; (g): Saprobic weight; (S): Saprobic score

References

An H, Jung G, Kim C (2012) Molecular taxonomy of a phantom midge species (*Chaoborus flavicans*) in Korea. Anim Syst Evol Divers 28(1):36–41

Bodescu F, Ion R, Cobzaru I, Ion M (2010) Inventarierea şi cartarea biodiversităţii din Rezervaţia Naturală Lacul Cuejdel–date preliminare. In: Instit Biol Buchaest–S Ann Com Sci (Abstract vol) 50(1):30–32

Grabowski A (1973) The biomass organic matter contents and calorific value of macrophytes in the lakes of the Szeszurpa drainage area. Pol Arch Hydrobiol 20(2):269–282

Mihu-Pintilie A, Romanescu G, Stoleriu C, Stoleriu O (2014) Ecological features and conservation proposal for the largest natural dam lake in the Romanian Carpathians-Cuejdel Lake. Int J Conserv Sci 5(2):243–252

Mititelu D, Monah F, Nechita N (1992) Contribuţii la studiul florei şi vegetaţiei higro-hidrofite din judeţul Neamţ. Stud Cerc Biol Muz Nat Sci-Piatra Neamţ 6:53–60

Munteanu D (2009) Păsări rare, vulnerabile și periclitate din România. Alma Mater Publishing, Cluj-Napoca

Orberdorfer E (1950) Beitrag zur Vegetationskunde des Allgäu. Beitr. Naturk Forsch Südw-Dtld Karlsruhe 9(2):29–98

Oltean M, Negrean G, Popescu A, Roman N, Dihoru G, Sanda V, Mihăescu S (1994) Lista roșie a plantelor superioare din România. In: Oltean M (ed) Studii, sinteze, documentaţii de ecologie, vol 1, 1st edn. Romanian Academy, Institutul de Biologie, Bucharest, pp 1–52

Oprea A, Sârbu C (2008) Rare plants in Stânișoarei mountains (Eastern Carpathians). J Plant Develop 15(1):47–61

Porumb MA (2006a) Cercetări privind algoflora planctonică a lacurilor de baraj din cursul mijlociu și inferior al râului Bistrița. Stud Cerc Muz Nat Sci Piatra Neamț 10:53–65

Porumb MA (2006b) Studiu privind algele din lacul de baraj natural Cuejdel-Neamt, Stud. și Cercet. Muz. Stud Cerc Muz Nat Sci Piatra Neamț 10:65–77

Resmeliță I, Pop I (1972) Considerații fitoceanologice asupra pajiștilor de Trifolia–Lolietum perennis Krippelova 1767, Resmeriță et al. 1967 din România. Contrib Bot–Cluj-Napoca 115–195

Sanda V, Popescu A (1997) Date privind cronologia, ecologia și cenologia speciei Typha laxmannii Lepechin din România. Stud Cerc Biol–S Biol Veget 49(1–2):67–75

Sanda V, Öllerer K, Burescu P (2008) Fitocenozele din România-Sintaxonomie, structură, dinamică și evoluție. ARS Docendi Publishing, Bucharest

Seghedin TG (1989) Noi asociații vegetale din Munții Bistriței. Ann Muz Suceava Coundy-Nat Sci 12:123–137

Specziár A (2008) Life history patterns of *Procladius choreus*, *Tanypus punctipennis* and *Chironomus balatonicus* in Lake Balaton. Ann Limnol-Int J Limnol 44(3):181–188

Tarmo T (2013) The genus *Potamothrix* (Annelida, Oligochaeta, Tubificidae): a literature review. Estonian J Ecology 62(2):121–136

Chapter 9
Sustainable Management of the Cuejdel Natural Dam Lake—Biodiversity Conservation Principles

Abstract This chapter aims to highlight the ecological features of this limnosystem that could further help to identify better solutions for the conservation of the lacustrine basin, in the context of an increasing anthropic pressure (deforestations, water pollution, uncontrolled tourist impact, etc). Cuejdel Lake was officially designated a National Natural Reserve by the Governmental Decision 2151/2004 regarding the creation of new protected areas and the conservation of natural flora and fauna habitats (IUCN IV.40 Cuejdel Lake), but this status is not enough for their protection. For this reason, it is considered that Cuejdel Lake area meets all the legislative requirements to establish a Site of Community Importance (SCI), a Special Area of Conservation (SAC) and/or a Special Protection Area (SPA), according to Natura 2000. The research indicates that the integration of the Cuejdel Natural Reserve and its catchment basin within the Vânători-Neamț Natural Park could be the best solution to sustainable management of this hydro-morphological site and its biodiversity. Also, the preservation of the aquatorium must represent a priority for the competent authorities and for the local population, directly involved in the conservation of the largest natural dam lake in Romania.

9.1 Management of the Lacustrine Water Body

The natural dam lakes in the mountain areas usually form as a result of geomorphological processes, with a strong impact on the landscape (landslides, muddy flows, thaws, floods). This feature determines a relative instability to the depression form behind the natural dam. Many lakes of this type might bring dramatic damage by flooding inhabited areas and especially by releasing a large quantity of water following the destruction of the fragile dams. There have been numerous such events around the world, the Eastern Carpathians included (Ichim et al. 1996; Rudoy 2002; Schneider et al. 2011; Romanescu et al. 2013).

Similar to the Red Lake in the Hășmaș Massif, the Cuejdel Lake is one of the few lacustrine basins formed by deluvial damming in the Eastern Carpathians area that still exists. However, the fluctuation levels that have occurred so far, despite its only 25 years of water accumulation, indicate a more intense hydrodynamic activity than

A. Mihu-Pintilie, *Natural Dam Lake Cuejdel in the Stânişoarei Mountains, Eastern Carpathians*, https://doi.org/10.1007/978-3-319-77213-4_9

that of its correspondent on the Bicaz Valley. It is for this reason that the water resource management plan ought to be a priority for the local authorities, as represented by the Siret Water Basin Administration.

In light of the results of this research, the sustainable management of the Cuejdel Lake needs to take into account several priorities. Perhaps the most important of these is to interdict forest exploitation in the catchment basin. Currently, out of the 10.38 km^2 of the catchment basin, 9.98 km^2 are occupied by deciduous, coniferous and mixed forests, while the remaining 0.4 km^2 are covered by subalpine meadows, river beds and bare land. Nevertheless, according to the data obtained by estimating the erosion in the catchment basin using both direct (the GPR technique) and indirect (USLE) methods, soil erosion ranges between 3.02 and 4.33 t/ha/year. Considering a 33% effluence rate, one can estimate a quantity of 3288.5–4521.26 m^3/year of eroded land that reaches the Cuejdel Lake (Fig. 9.1).

According to calculations estimating the silting rate, the Cuejdel Lake will naturally silt up in about 250 years if the current vegetation conditions remain and unless erosion in the catchment basin area intensifies. If the areas covered by forests were reduced by 50%, the lifespan of the lake would diminish by 75–100 years. In another scenario, where the entire water basin were to be deforested, the lake would silt up in 75 to 100 years, given an accelerated soil input of >22,000 m^3/year. These values clearly indicate the role that forests play in preserving the water body (Fig. 9.2).

Fig. 9.1 Accumulation of terrigenous deposits at the mouth of Cuejdel and Glodu creeks resulted from deforested areas upstream of the lake

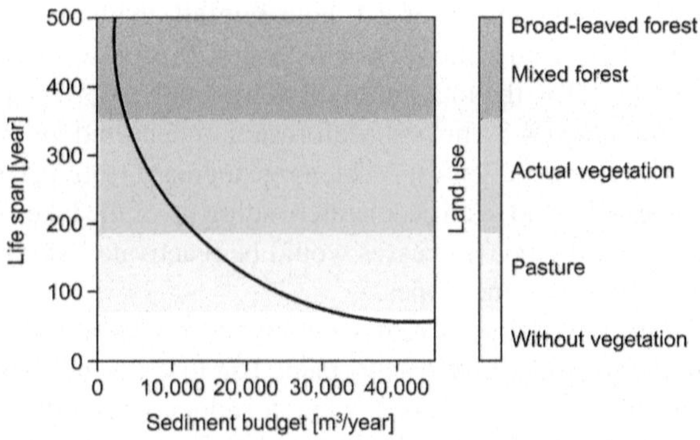

Fig. 9.2 Correlation between the lifespan of Cuejdel lake and the increasing of sediment budget determined by changing of land use in catchment lacustrine basin

Table 9.1 Forest cover degree influence [%] on maximum flow stream probability [Q_{max}] with insurance of 5% [m^3/s] in the catchment basin of Cuejdel lake calculated based on rational method (Diaconu 1992; Mihu-Pintilie et al. 2016)

Forest cover degree [%]	[Q_{max}] with insurance of 5% [m^3/s]					
	Cuejdel	Glodu	Făgetu	Piciorul Crucii	Rotund brook	Total lake basin
80–100	1.81	1.37	0.64	0.33	0.20	4.35
60–80	1.89	1.42	0.67	0.35	0.21	4.54
40–60	1.96	1.48	0.69	0.37	0.22	4.72
20–40	2.04	1.54	0.72	0.38	0.22	4.90
0–20	2.11	1.59	0.74	0.40	0.23	5.07

To the same extent, the forest roads have a negative impact on the slopes around the lake. Considering that the landslide under the Muncelul Peak was triggered by a forest road and eventually led to the forming of the Cuejdel Lake, it is just as possible to reactivate the landslide activity in the same manner once again. All throughout the time our research was conducted, large logging trucks were observed in the upper area of the lake, especially in winter.

Along with reducing the forested areas, the flow of the lake's tributaries would increase exponentially, especially during heavy rainfall. According to the rational formula of determining the maximum discharge of 5% probability on <10 km^2 water basins as formulated by Diaconu (1992), maximum river discharges of 0.2–1.81 m^3/s (Cuejdel—$Q_{5\%}$ 1.81 m^3/s; Glodu—$Q_{5\%}$ 1.13 m^3/s; Făgetu—$Q_{5\%}$ 0.64 m^3/s; Piciorul Crucii—$Q_{5\%}$ 0.33 m^3/s; Rotund Brook—$Q_{5\%}$ 0.2 m^3/s;) have occurred in the past 20 years. Given the 0.67 circularity coefficient of the catchment basin and the fact that the time of concentration of rainfall on each sub-basin is less than 70 s, we consider that a discharge of approximately $Q_{5\%}$ 4.37 m^3/s flowed almost simultaneously into the lake at least once in the past 20 years. If the duration of such flood were to be maintained for about 30 min, a 7866 m^3 water volume would reach the lake, enough to raise its level by 5.36 cm (Table 9.1).

In the hypothesis of a 1–hour rainfall, the water level would have increased by 10.72 cm. Under these circumstances, if each water sub-basin reduced its forested area by 50%, the maximum discharge with a 5% overflow probability would reach a cumulated 4.87 m^3/s. If deforestation occurred entirely, the maximum $Q_{5\%}$ value would reach 5.07 m^3/s. A discharge increase would also intensify erosion and would eventually lead to an accelerated silting up of the lake. At the same time, the geomorphological slope processes would be reactivated along the forest exploitation roads or on high-incline slopes.

Such hydrologic events with higher or lower intensity have occurred at least five times since the lake was formed. The first among these, mentioned by Ioniță et al. (1996), occurred in 1992, when a sudden flood determined a dramatic increase of the water level, which led to breakage into the yet unconsolidated dam. The event led to an abrupt level drop of 60 cm. From that event onwards, exceptional flood also occurred on: 28.VII.1999, 05.VIII.2004, 12.VII.2005, 26.VI.2010 and 12.IX.2012. The impact of these weather events triggered a further 2 m drop of the lake level, the deepening of the discharge outlet on the dam being the main cause (Fig. 9.3).

Based on the probable data of the discharge and its impact on level fluctuation, a series of hydrotechnical measures could be recommended to preserve the Cuejdel Lake. Perhaps the most necessary of all would be to build a dam in the discharge outlet area, where erosion accelerates water loss. This intervention can be facilitated by the solid geologic structures at the bottom, but especially by the length of the water bed, which is no wider than 10–15 m in the upper shore area (Fig. 9.3).

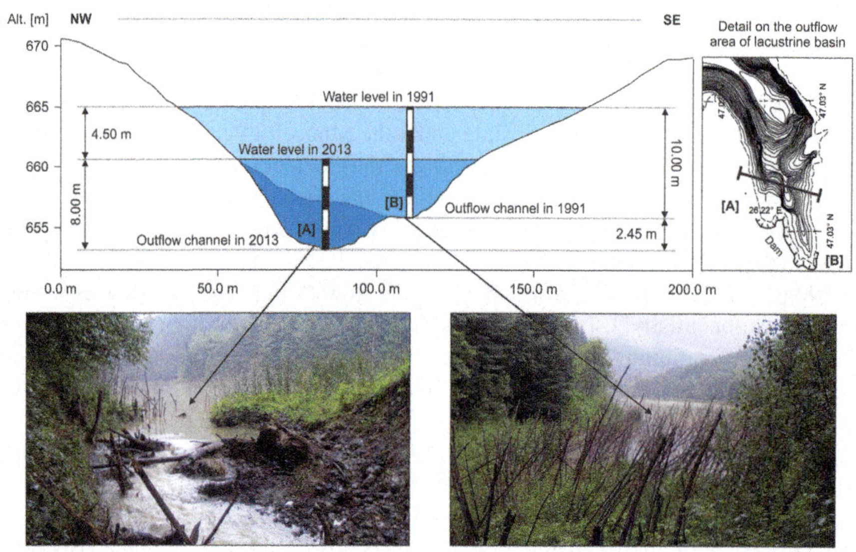

Fig. 9.3 Changing the outflow of lake mainly due to floods. In the left photo is current drainage channel and in the right photo it is the initial drainage channel, currently inactive

Considering the intense torrential rainfall in the area, several technical aspects such as building a dam should be taken into consideration, as breakage during high water level rises might occur. One such example is the dam lake Iezerul Sadovei in the Feredeu Ridge, where as a result of the progressive releases of water, the local authorities built a concrete dam with an overflow installation for the excessive drainage. This measure proved to be totally ineffective, as the concrete structure was fitted only in the friable deluvial mass. After only a few years, during which many flood episodes occurred at the catchment level, the water generated a breach around the anthropic dam rendering the structure useless.

Another useful measure that could extend the warping period of the lake could be the installing of sediment nets on the Cuejdel, Glodu and Făgetu creeks. Currently, on Cuejdel creek, approximately 100 m upstream from the lake, the bridge that connects the forest road with Magazia to the left bank acts as such a net. Here, behind the concrete pillars, rough alluvial shoals appeared, mainly from gravel, galets and wooden materials; nevertheless, the fine solide charge of the creek is easily transported trough the active section of the bridge and deposited further away in the end tail of the lake. In the case of Red Lake, installing anti-erosion protection dams on the Oaia and Suhard tributaries considerably reduced the ratio of alluvial effluence (Romanescu et al. 2013).

As a conclusion, we believe that the main conservational measure of the natural dam Cuejdel Lake should include the following aspects related to sustainable management of the water body:

- Interdiction of forest explotation at the level of catchments and the strict control of frequent illegal cutting in the area.
- Protecting the left bank through interdiction of heavy tonnage traffic on the forest road that connects Magazia to the downstream area of the lake. The upstream barrier is not functional.
- Regular cleaning of drains on the Făgetu creek and its tributary on the right, where the erosion of lacustrine banks is the most intense.
- Building a dam in the area of the natural barrage in order to prevent the deepening of the drainage channel and constant water losses.
- Installing sediment nets on the Glodu and Cuejdel tributaries, to catch the alluvial material upstream the lake and reduce the warping ratio.
- Placing a hydrometric rod and establishing a permanent supervision program of Cuejdel water body, with daily readings of levels and evaluation of water balance (input-output).
- Elaborating a management plan for emergency situations (overflows, floods, etc.) and management water volumes through manipulation of the dam.
- Interdiction of any kind of activities that can determine the contamination with pollutants (hydrocarbons, plastic waste, sawdust, etc.) and affect the aquatic biodiversity.

9.2 Protected Area Status—IUCN IV.40 Cuejdel Lake

Cuejdel presents, alongside multiple economic interests, an extraordinary scientific potential. Its formation allows the specialists, especially geographers and biologists, a better, more subtle understanding of the genesis, evolution and impact that natural barrage lakes from the Carpathian area of Romania have over the landscape. At the same time, the aquatic and paludous biocoenoses developed around such an aquatorium leads to the creation of new habitats and improvement of biodiversity. That is why the public administration should be directly interested in preservation and management of the lake for a sustainable administration (Mihu-Pintilie et al. 2014; Stoleriu et al. 2014).

Even if Cuejdel Lake exists from 1991, only in 2004 it was included on the list of protected areas from Romania according to Government decision No 2151/2004 regarding the establishment of protected area for new regions (IUCN IV.40 Cuejdel Lake). Out of the 74 natural reservations established that year, 9 being lakes (for example—IV.4. Bârcă's Lake; IV.6. Bălătău Lake; IV.28 Bugeac Lake; IV.30. Vederoasa Lake, etc.), "Cuejdel Lake Natural Reserve" (IV.40.) belongs to the IV IUCN category (International Union for Conservation of Nature) (Fig. 9.4).

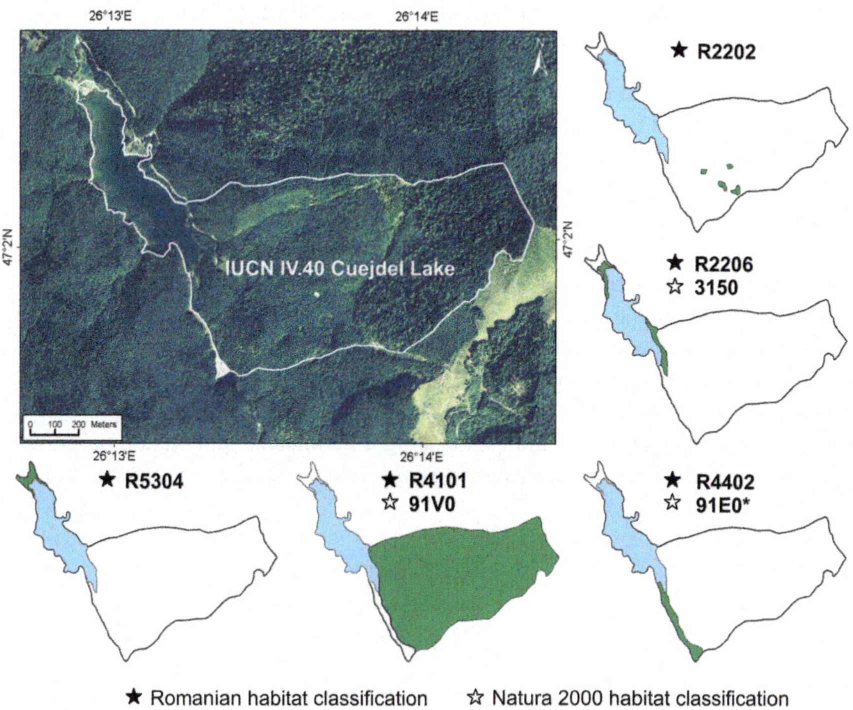

Fig. 9.4 Distribution of habitats within the Cuejdel natural reserve (Mihu-Pintilie et al. 2014)

The protected area surrounding Cuejdel Lake covers a surface of 114 ha from the administrative area of Cuejdiu locality (Gârcina Commune, Neamţ County) and includes within its limits, the water body and the landslide surface that formed the natural dam. The surrounding forest parcels and sub-parcels are: 113T for Cuejdel Lake, 41A, 41B, 42A, 42B for the landslide area included in UP II Mijlocul (Gârcina Forest District). Mainly, this type of habitat and species management is focused on preservation with possibility of exploitation interventions.

According to Emergency Ordinance 57/2007 regarding the protected areas regime, preservation of natural habitats, wild flora and fauna, approved with additions and interventions through Law No 49/2011, these type of natural reservations are protected areas developed in order to protect and preserve some important natural habitats and species—such as fauna, floristic, forests, hydrological, geological, speleologic, paleontological, pedological habitats. In spite of these regulations, in the "Cuejdel Lake Natural Reserve" anthropic activities related to tourism, education, scientific research, fishing, control of natural risks and administration of forest resources are permitted.

The general legislative frame concerning the preservation and administration of the Cuejdel Lake Protected Area (IUCN IV.40) includes 40 Government decisions, Laws and Emergency Ordinances. Still, even under their incidence, many illegal activities take place within its limits. For example, the intensive fishing almost led to extinction of the indigenous trout (*Salmo trutta fario* L), and also to appearance of species with negative impact on the fish communities diversity. One of these is the perch (*Perca fluvialitis* L.), a predatory fish that consumes also the spawns that ensures the density of fish communities. Another very important problem is represented by the household waste dumped on the lake shores. Fortunately, the low accesibility of the lake and the heavily forested shores determine a relatively low touristic activity.

Even if the administrative formalities were finalized in 2004, establishing scientific studies for the Management Plan of Cuejdel Lake began only in 2010 (PM v1 from 15.II.2012) and were financed by the Ministry of Environment and Climate Changes with support from the European Union. According to these studies, within the natural area were recorded 38 species of birds, 8 species of amphibians, 2 species of reptiles and 11 habitats were identified. Out of these, 5 habitats, 8 species of birds, 3 species of amphibians and one reptile species are mentioned in the modified version from 2006 (2006/105/CE) of the Habitat Directive (92/43/CEE) (Table 9.2) and Birds Directive (79/409/CEE) (Table 9.3).

Although a series of important species and habitats were signaled, because the legislation states the necessity of establishing a Site of Community Importance (SCI), or a Special Area of Conservation (SAC) and/or a Special Protection Area (SPA), the national authorities did not adopt any of these measures yet. This is the reason why the Cuejdel Lake and its surrounding area are not included within the SCI and SPA limits (vectorial format shapefile), published in 2011 on the Ministry of Environment and Climate Changes website. We believe this measure is unavoidable, as two other sites with community importance (ROSCI0270, ROSPA0107) were established only 1.7 km north of the lake limit. According to our research results, "Cuejdel Lake

Table 9.2 Habitats indentified in the Cuejdel natural reserve, according to the Romanian and Natura 2000 classifications (Mihu-Pintilie et al. 2014)

Romanian classification (Doniță et al. 2005)	Natura 2000 habitat classification	International regulations
R2202: Danubian communities with *Lemna minor, L. trisulca, Spirodela polyrhiza* and *Wolffia arrhiza*	–	–
R2206: Danubian communities with *Potamogeton perfoliatus, P. gramineus, P. lucens, Elodea canadensis* and *Najas marin*	3150 Natural eutrophic lakes with *Magnopotamion* or *Hydro-charition*	Habitats directive (92/43/EEC) modif. 2006/105/CE
R5304: Danubian communities with *Sparganium erectum, Berula erecta* and *Sium latifolium*	–	–
R4101: South-east Carpathian forests with spruce *(Picea abies)*, beech *(Fagus sylvatica)* and fir *(Abies alba)* with *Pulmonaria rubra*	91V0 Dacian beech forest (Symphyto-Fagion)	Habitats directive (92/43/EEC) modif. 2006/105/CE
R4402: Dacian-getic forest on hilly meadows with black alder *(Alnus glutinosa)* and *Stellaria nemorum*	91E0*Alluvial forest with *Alnus glutinosa* and *Fraxinus excelsior* (*Alno-Padion, Alnion incanae, Salicion albae*)	Habitats directive (92/43/EEC) modif. 2006/105/CE

Natural Reserve" meets all the criteria specified in the EU Directives and needs to be included in the European Network Natura 2000 (Table 9.4).

For a better preservation and sustainable revaluation of Cuejdel Lake, another viable solution could be its integration within the boundaries of Vânători Neamț Natural Park. The mentioned area is already a good example of sustainable management for the protected areas in Romania and presents a good usage of European funds. Therefore, through the extention of the southern limit with approximately 3 km, the benefits and stricter internal measures of areas from Natura 2000 network could significantly improve the conservation status of Cuejdel Lake.

Moreover, within the limits of Vânători Neamț Natural Park there is a consistent population of aurochs (*Bison bonasus*) raised in semi-captivity. In 2012, five aurochs were released in the Neamț Mountains area, as a first step in the efforts of the park's administration to repopulate the Carpathian forests with this species. In the same context, the proximity to Cuejdel Lake can be used to enlarge the natural protected

Table 9.3 Species from Cuejdel natural reserve listed in the birds and habitats directives (Mihu-Pintilie et al. 2014)

Species names	Type	Birds directive (79/409/EEC), habitats directive (92/43/EEC) modif. 2006/105/CE
Picus canus	Bird	Annex 1 bird
Turdus merula	Bird	Annex 1 bird
Turdus philomelos	Bird	Annex 1 bird
Bonasa bonasia	Bird	Annex 1 bird
Columba palumbus	Bird	Annex 1 bird
Dendrocopos syriacus	Bird	Annex 1 bird
Garrulus glandarius	Bird	Annex 2 bird
Anas platyrhynchos	Bird	Annex 2 bird
Bombina variegata	Amphibian	Annex 2 habitat
Rana temporaria	Amphibian	Annex 5 habitat
Triturus cristatus	Amphibian	Annex 2 habitat
Lacerta agilis	Reptile	Annex 4 habitat

Fig. 9.5 Birdwatching tower and tourist information panels around the protected area of Cuejdel lake

area ROSCI0270 of Vânători Neamț Natural Park, a measure that could be useful for a future extension of the space where aurochs can adapt to freedom (Fig. 9.5).

Table 9.4 Legal frameworks concerning the conservation and management of Cuejdel lake protected area (Mihu-Pintilie et al. 2014)

Legal frameworks
• Ordinance No. 195/2005 on environmental protection, approved by Law 265/2006, with subsequent amendments
• Code For Forest - Law No. 46/2008, with subsequent modifications
• O.U.G. No. 85/2006 on establishing detailed assessment of forest damage and forest vegetation outside their
• Law No. 31/2000 on the establishment and sanctioning contraventions forest, with subsequent modifications
• H.G. No. 1076/2009, for approval of forest guards
• O.U.G. No. 23/2008, approved by Law 317/2009, respect fisheries and aquaculture, with subsequent modifications
• Law No. 407/2006 on the protection of hunting and hunting fund, with subsequent modifications
• Law No. 13/1998 for Romania's accession to the Convention on the conservation of wildlife and natural habitats in Europe, adopted at Berne on September 19, 1979
• Law No. 58/1994 for ratification of the Convention on Biological Diversity signed in Rio de Janeiro on June 5, 1992
• Law No. 13/1998 for Romania's accession to the Convention on Conservation of Migratory Species of Wild Animals, adopted in Bonn on June 23, 1979
• Ministry of Agriculture and Rural Development Order No. 353/2008 approving the Regulation on licensing, organization and hunting
• Ministry of Environment and Forests Order No. 19/2010, for approval of the Methodological Guide for proper evaluation of potential effects of plans or projects on natural protected areas of interest
• O.U.G. No. 57/2007 on the regime of natural protected areas, natural habitats, flora and fauna, with subsequent modifications
• 107/1996 Water Act, with subsequent modifications
• Environment and Sustainable Development Ministerial Order No. 662/2006 approving the Procedure and powers of issuing permits and water management
• Law No. 5/2000 concerning the approval of national planning - Section-III protected areas
• Law No. 5/1991 for Romania's accession to the Convention on Wetlands of International Importance especially as Waterfowl Habitat (February 2, 1971 Convention on Wetlands of International Importance especially as Waterfowl Habitat Ramsar, as was amended by the Protocol of Paris on December 3, 1982)
• Environment and Sustainable Development Ministerial Order No. 1533/2008 on approving the methodology for the award of administration of natural protected areas that require the creation of management structures and methodology for awarding custody of protected natural areas that do not require the creation of administrative structures
• Environment and Sustainable Development Ministerial Order No. 1964/2007 concerning the establishment of protected area system of sites of Community importance, as part of European ecological network Natura 2000 in Romania

(continued)

Table 9.4 (continued)

Legal frameworks
• H.G. No. 1284/2007 declaring Bird specially protected areas as part of the European ecological network Natura 2000 in Romania
• Environment and Sustainable Development Ministerial Order No. 1338/2008, regarding procedures for issuing opinion Natura 2000
• H.G. No. 1076/2004, establishing a procedure for carrying out environmental assessment for plans and programs
• Law No. 350/2001 of planning and urban development, with subsequent modifications
• Law No. 315/2004 on regional development in Romania, with subsequent modifications
• Law No. 481/2004 on civil protection, republished in 2008
• H.G. No. 1143/2007 concerning the establishment of new protected areas
• Environment and Sustainable Development Ministerial Order No. 1710/2007 approving the necessary documentation to establish the system of protected area of national interest
• Environment and Sustainable Development Ministerial Order no. 410/2008 approving the authorization procedure for harvesting, capture and/ or acquisition and/or marketing on domestic or export of flowers hand, the fossils of plants and vertebrate and invertebrate fossils, and flora plants and animals, respectively wildlife and their importation
• H.G. No. 1635/2009 on the organization and functioning of the Ministry of Environment and Forests
• H.G. No. 459/2005 on the reorganization and operation of the National Agency for Environment, with subsequent modifications
• H.G. No. 112/2009 on the organization and functioning of National Environmental Guard, with subsequent modifications
• H.G. No. 1194/2008 on the organization, structure and functioning of the National Agency for Fisheries and Aquaculture, with subsequent modifications
• H.G. No. 229/2009 on the reorganization of the National Forests - Romsilva and approval of the organization and functioning
• H.G. No. 333/2005 for the reorganization of territorial divisions of forestry and hunting regime territorial inspectorates forestry and hunting regime
• Order No. 756/1997 approving the regulation on the assessment of environmental pollution, with subsequent amendments
• O.U.G. No. 68/2007 on environmental liability with regard to preventing and remedying environmental damage, approved by Law 19/2008, with subsequent modifications
• H.G. No. 188/2002 approving the rules on conditions for discharge of wastewater into the aquatic environment, with subsequent amendments
• O.U.G. No. 78/2000 concerning, approved by Law 426/2001, with subsequent amendments
• Law No. 86/2000, ratifying the Convention on Access to Information, Public Participation in Decision Making and Access to Justice in Environmental Matters, signed at Aarhus on 25/06/1998
• H.G. No. 878/2005 on public access to environmental information, with subsequent modifications

Starting with 2014, within the Operational Program POS Environment—Priority Axis 4, the ongoing project entitled: "Studies concerning the analysis of marking the natural limits of natural protected areas from the perspective of elements with conservation interest for the areas already declared natural protected area" was developed. As a member in the implementation team from SOR (Romanian Ornithological Society), I have presented the documentation supporting the proposal for extending the limits of area with conservational interests IUCN IV.40 Cuejdel Lake. Still, it has to be said that it is not an easy process, redrawing the limits of a natural protected area triggering multiple legislative and economical implications.

References

Diaconu C (1992) Probleme de metodica determinării curbelor teoretice ale probabilităţilor de depăşire (asigurare) în hidrologie. Hidrotehnica–Extras, 37(4–5) (Bucharest)

Ichim I, Rădoane N, Rădoane M (1996) Procese geomorfologice cu intervale de recurenţă mare în arealul munţilor flişului. Exemplificări din Judeţul Neamţ (Geomorphological process with high frequences in flysch mountains area. Examples from Neamț County). In: Studii şi cercetări—extras 8, Muzeul de Ştiinţe Naturale Publishing, Piatra-Neamţ, p 15

Mihu-Pintilie A, Romanescu G, Stoleriu C, Stoleriu O (2014) Ecological features and conservation proposal for the largest natural dam lake in the Romanian Carpathians-Cuejdel lake. Int J Conserv Sci 5(2):243–252

Mihu-Pintilie A, Asăndulesei A, Nicu IC, Stoleriu CC, Romanescu G (2016) Using GPR for assessing the volume of sediments from the largest natural dam lake of the Eastern Carpathians: Cuejdel lake. Romania. Environ Earth Sci 75:710. https://doi.org/10.1007/s12665-016-5537-1

Romanescu G, Stoleriu C, Enea A (2013) Limnology of the Red lake, Romania. An interdisciplinary study. Springer, Dordrecht, Heidelberg, New York, London

Rudoy AN (2002) Late-Quaternary diluvial floodstreams in the mountains of Altai and Tuva–paleoglaciological base, geological effect, level of problem study. Lunar Planet Sci 33:1109–1110

Schneider JF, Gruber FE, Mergili M (2011) Recent cases and geomorphic evidence of landslide—dammed lakes and related hazards in the mountains of Central Asia, In: Proceedings of the second world landslide forum, vol 1, issue 1–7. Rome, 3–9 October 2011

Stoleriu CC, Stoleriu O, Mihu-Pintilie A (2014) Scientific and tourist value of natural dam lakes in the Carpathian Mountains (Romania). Case study: Red, Cuejdel and Iezerul Sadovei lakes. In: SGEM–Proceedings of the 14th international multidisciplinary scientific geoconferences–ecology and environmental protection, vol 14, issue 2. Albena, 18–24 June, pp 625–632

Chapter 10
Conclusions

Abstract On the territory of Romania there is a relatively large number of water units that meet the conditions mentioned within the universally accepted definition of lakes. Even so, the lacustrine basins formed by natural processes are rare and they are often modified by human activities. The natural dam lakes correspond to a special genetic category, due to the processes and phenomena that include them in the landscape. The best represented category on the territory of Romania is the one of natural dam lakes formed due to landfall and landslide processes. The most favorable conditions for their formation are found in the Eastern Carpathians, Curvature and Moldavian Subcarpathians, but also in some isolated areas from the Moldavian Plateau. The natural dam lake Cuejdel, also known by the hydronym Cross Lake, situated in the south-east part of Stânişoarei Mountains (NE Romania), is registered in the geographic literature as one of the newest natural aquatorium (25 years old). The results of the integrated monitoring program on the lacustrine basin and the surrounding area, based on the investigations related to the biotope and biocoenoses, indicates the fact that Cuejdel Lake describes an aquatic entity with unique elements in the landscape of Eastern Carpathians.

On the territory of Romania there is a relatively large number of water units that meet the conditions mentioned within the universally accepted definition of lakes. Even so, the lacustrine basins formed by natural processes are rare and they are often modified by human activities. The highest frequency of natural lakes occurs in the area of Carpathian Mountains and Subcarpathia, in the Danube Delta and rarely in the plateau and plain area.

The natural dam lakes correspond to a special genetic category, due to the processes and phenomena that include them in the landscape. These are often considered a consequence of hydro-geomorphological processes, and their formation in the inhabited areas could represent a risk factor for human society. Around the world, based on the genetic category and nature of the obstructive dam, the following types have been identified: volcanic dam lakes (Nicaragua, Snag, Rio Magdalena), lakes formed by landslide (Earthquake, San Cristobal, Spirit, Yamaska, Addergoogle Bog, Goatswater), moraine lakes and ice dams (Gapshan-Shyok, Nostetuko, Rio Plomo), fluviatile dam lakes (Pepin, Tung-ting, Tulare, Lacul Albas-

A. Mihu-Pintilie, *Natural Dam Lake Cuejdel in the Stânişoarei Mountains, Eastern Carpathians*, https://doi.org/10.1007/978-3-319-77213-4_10

tru—California, Old River), aeolian dam lakes (Moses), former lagoons blocked by barrier beach (Eureka, Razim-Sinoe), organic dam lakes or formed by animals (Okeechobee, Beaver).

The best represented category on the territory of Romania is the one of natural dam lakes formed due to landfall and landslide processes. The most favorable conditions for their formation are found in the Eastern Carpathians, Curvature and Moldavian Subcarpathians, but also in some isolated areas from the Moldavian Plateau. The highest susceptibility is held by the Carpathian flysch area where the formation of lacustrine accretions behind the diluvium is frequent. The most popular examples are: Red Lake (Hășmaș Mountains—Bicaz), Mocearu Lake (Buzău Mountains—Buzău), Bălătău Lake (the Curvature Carpathians—Zăbala), Balătău Lake (Ciuc Mountains—Trotuș), Bețis Lake (Maramureș Mountains—Ampoi), Iezer and Bolătău Lakes (Feredeu Ridge—Sadova), Tăul Zânelor Lake (Bârgău Mountains—Colibița), Izvorul Măgurii Lake (Bârgâu Mountains—Ilva), Devil's Lake (Nera Canyon—Nera).

The natural dam lake Cuejdel, also known by the hydronym Cross Lake, situated in the south-east part of Stânișoarei Mountains, component of the Central Group within Eastern Carpathians, is registered in the literature as one of the newest natural aquatorium (25 years old). The present lacustrine basin has been forming in the upper basin of Cuejdiu River (left tributary of Bistrița River) at about 1.5 km upstream Cuejdel brook (left tributary of Cuejdiu River). Its origin is based on a large landslide (surface: 67.74 ha) that produced the rock displacement during the summer of 1991, on the left flank of Muncelu Peak (a subunit of Neamț Mountains, relative altitude of 1077 m).

The analysis of physico-geographical conditions from Cuejdel Valley and their role in the formation of Cujdel Lake emphasize a number of factors that led to the destabilization of the versants and generated the slide complex. It is important to mention the role of: geological factors (flysch layers with clay and argillaceous-marl deposits and compact sandstones), relief (relief energy >300 m, declivity 30°–50°), rainfall regime (cumulated pp. 05–07. 1991—741.4 mm), hydrography (Order IV—Horton Strahler classification) and the structure of valleys (transverse/longitudinal on the general direction of lithologic layers), bioedaphic cover (deciduous, coniferous and mixed forests—96.2%), and last but not least, the anthropic factor (the density of forest exploitation road network 10.67 km/km^2). The list of factors that led to landslides can be completed by the earthquake from 1990 (5.4°Richter).

The first measurements in the lacustrine basin of Cuejdel Lake were taken in 2011 using a high precision echosounder (Sonar Valeport Midas Ecosounder) and a GPS for topographic correction (GPS Leica System 1200). Morphographic and morphometric parameters determined during the same year were indicating the following values: 661 m for altitude of water surface, 13.95 ha for surface of lacustrine bottom, 1.004 km max length, 19.25 m min width, 138.8 m med width, 282.6 m max width, perimeter of 2.8 km, 16.45 m max depth, 5.6 m med depth, volume of 92.53×10^4 m^3. Based on the results, Cuejdel Lake confirms the superlative of the largest natural dam lake in Romania, exceeding in terms of surface and maximum depth, the dimensions of Red Lake from Bicaz Valley (surface of 12.1 ha, 10.5 m max depth).

The second stage of measurements was taken during the winter of 2013, this time using GPR technology (Ground Penetrating Radar) with the help of GPR Malå Ramac X3 M device and a 100 MHz antenna. Because this method of bathymetric scanning of alluvial deposits requires an outright plan and also a frozen surface, the dimensions of morphometric parameters of the lake have been reduced compared to previous ones, but normal for the cold season. The results of bathymetric scanning were concretized by the reconstruction of the initial basin and the achievement of a map showing the thickness of lacustrine sediments. Therefore, the aquatorium dimensions during 1991 have reached the following maximum values: 665 m for altitude of water surface, 16.22 ha for surface of lacustrine bottom, 1.17 km max lenght, 24.1 m min width, 138.6 m med width, 341.2 m max width, perimeter of 2.95 km, 18.8 m max depth, 7.5 m med depth, volume of 122.3×10^4 m^3. The values were confirmed both by the internal structures identified on GPR profiles interface and the limit of maximum levels left on the arbors found in the water mass.

The regressive evolution of lake levels that took place during the 25 years, indicates a decrease of ca. 3.5–4 m, that determines the medium withdraw of the surface with 0.1 ha/year. The main cause is represented by the sinking of the drainage channel from the obstructive dam surface, due to flash floods subdued by the lake. The lake volume presents a descending tendency, being reduced per total with 297,654.77 m^3, from which, according to GPR measurements, 103,988.87 m^3 were replaced by alluvial deposits transported in lacustrine depression by the tributaries.

The map showing the distribution of sediment thickness indicates a discrepancy between 0.1 and 3.6 m, from which the interval 0.1–1.0 m covers 8.552 ha (60.7%), the interval 1.0–2.0 m covers 4.676 ha (33.32%), and the interval >2.01 m covers the rest of 8326.5 m^2 (5.93%). The sedimentation rate is included on different sectors, between 1 and 16 cm/year, and the minimum rate of accumulation for terrigenous sediments is situated around the medium value of 4521.26 m^3/year. This value, related to the surface of reception lake basin (8.94 km^2) and to an effluence rate of 33% determined by the level of vegetation cover, indicates a medium production of sediments in the basin—cca. 4.33 t/ha/year. Therefore, it can be estimated that the natural dam lake Cuejdel will clog, maintaining its hydro-geomorphological balance in 247 years from now.

The monitoring program of Cuejdel Lake also included an analysis of biotic conditions, concerning the description of physico-chemical parameters (T°, pH, LDO101, CDC401), water quality index (WQI), trophicity level determined by chemism and nutrients (DO_2, CBO_5. $KMnO_4$, NO_3^-, NO_2^-, NH_4^+, PO_4^{3-}, P total). In situ measurements were taken every season (4 times/year), using a Hach Lange type Multiparameter with a real-time display, and the investigations from the laboratory were based on standard methods of water analysis.

Thus, in terms of temperature, Cuejdel Lake is acting as a dimictic aquatorium, with direct thermal stratification in the summer (T°—epilimn. 22°–23°; hipolimn. 4°–5°), spring (T°—epilimn. 16°–17°; hipolimn. 4°–5°) and autumn (T°—epilimn. 12.5°–13°; hipolimn. 4°–5°), and reversed during winter (T°—epilimn. 0°–1°;

hipolimn. 4°–5°). The amount of dissolved oxygen (LDO101) drops vertically and inversely proportional with the standard conductivity of the water (CDC401), indicating an euxinic stratification, induced by the absence of vertical currents. The pH value follows the same tendency on bathymetric levels, presenting a lightly alkaline environment, specific to mountain waters (med. pH 0–8.3). Due to these physico-chemical conditions, completed by the medium transparency of 1.8–2.0 m, Cuejdel Lake presents a productive environment only in the epilimnion zone (0–3 m).

Water quality index (WQI), determined based on the nutrients: nitrates (NO_3^-), phosphates (PO_4^-), ammonium ion concentration (NH_4^+), completed by the amount of dissolved oxygen (DO_2), biochemical oxygen consumption (CBO_5) and the pH, indicates a score of 86%, this value being equivalent for 1st quality class. Also based on the nutrients determined in laboratory samples, Cuejdel Lake belongs to mesotrophic aquatorium category, where the trophicity conditions resemble the ones from Sfânta Ana Lake, located in a volcanic crater or the ones from Brădișor and Izvorul Muntelui anthropic accretions.

The determination of trophicity level in aquatic limnosystem Cuejdel was based on ecological bio-indices found in the water mass. The interdisciplinary research in collaboration with biology experts was concretized by identifying the following categories: algal communities (phytoplankton—cca. 50 taxons), vertebrate aquatic fauna (cca. 17 species) and invertebrate aquatic fauna (over 150 taxons), from which is worth mentioning the insects (over 100 taxons), zooplanktonic communities (cca. 14 taxons) and benthic macroinvertebrates (cca. 7 taxons). Of all the species found, some with a higher density and others very rare, 72 taxons have shown, due to dominant or specific valence, an indicating saprobic value. Therefore, based on ecological bio-indices from the water mass, Cuejdel Lake is defined as an oligotrophic aquatic system with mesotrophic tendencies in the warm season.

The transition from a simple river sector specific to the Eastern Carpathians to an lacustrine basin on this scale, has determined, as expected, an increase of biodiversity. Thus, have been identified 11 aquatic and palustrine phytocoenoses that represent propitious habitats both for the existing species and the new ones. The identification of nine bird species, three amphibian species and one reptile listed in the Habitats Directive (92/43/CEE) and the Birds Directive (79/409/CEE), indicates the fact that a protected area status (IUCN IV.40 Cuejdel Lake) is not enough for their protection. For this reason, it is considered that Cuejdel Lake area meets all the legislative requirements to establish a Site of Community Importance (SCI), a Special Area of Conservation (SAC) and a Special Protection Area (SPA).

The final conclusion of the present study, achieved based on the investigations related to the description of biotope (lacustrine basin—sediments—water) and the aquatic and palustrine biocoenoses, indicates the fact that Cuejdel Lake describes an aquatic entity with unique elements in the landscape of Stânişoarei Mountains. The preservation of the aquatorium must represent a priority for the competent authorities (City Hall of Gârcina commune, Neamţ County Council, Romanian Waters National Administration "Siret"—Water Management System of Neamţ, ROMSILVA National Forest Administration, Ministry of Environment and Climate

Change) and also for the local population, directly involved in the conservation of the largest natural dam lake in Romania.

Future research directions. The results of the integrated monitoring program on the natural dam lake Cuejdel, acquired during four years of study at the "Doctoral School Chemistry and Life Sciences" within the Faculty of Geography and Geology, Department of Geography—Alexandru Ioan Cuza University of Iaşi (Romania) represent an intermediary stage in developing an important topic as the formation of natural dam lakes. In the future, the aim is to extend the limnogeographical analysis through a more active participation of specialists from fields related to limnology. This aspiration is a result of the necessity to realize an interdisciplinary model of investigation for aquatic and palestine environments, using the latest generation equipments and techniques.

Printed by Printforce, the Netherlands